FUELS, LUBRICANTS COOLANTS AND FILTERS

A training guide to the "hows" and "whys" of modern fuels, lubricants, coolants and filters

F U N D A M E N T A L S   O F   S E R V I C E

FUNDAMENTALS OF SERVICE (FOS) is a series of manuals created by Deere & Company. Each book in the series is conceived, researched, outlined, edited, and published by Deere & Company, John Deere Publishing. Authors are selected to provide a basic technical manuscript that is edited and rewritten by staff editors.

PUBLISHER
DEERE & COMPANY, JOHN DEERE PUBLISHING
ALMON-TIAC Building, Suite 140, 1300 - 19th Street,
East Moline, IL 61244
http://www.deere.com/deerecom/Farmers+and+Ranchers/Publications
1-800-522-7448

JOHN DEERE PUBLISHING STAFF
Cindy S. Calloway

TO THE READER: The main purpose of this manual is to give you the how's and why's of modern fuels, lubricants, coolants and filters. For the novice, it is a training guide; for the journeyman, a reference. The story is written in simple form using many illustrations so that it can be easily understood.

 CAUTION: Observe safety messages and information in this manual.

ACKNOWLEDGEMENTS: John Deere Publishing gratefully acknowledges help from the following groups: American Association for Agricultural Engineering and Vocational Agriculture; American Petroleum Institute; American Society of Agricultural Engineers (ASAE); Borg-Warner Corp.; Dana Corp.; Ethyl Corp.; Federal-Mogul Corp.; Gulf Oil Corp.; Imperial Oil Ltd.; National Board of Fire Underwriters; National Fire Protection Assoc.; National LP-Gas Association; Purdue University Extension Service; Shell Oil Co.; Society of Automotive Engineers; Standard Oil Co.; Texaco, Inc.; TRW Replacement Division.; Union Carbide Corp.; University of Missouri Experiment Station. Thanks also to a host of John Deere people for their valuable suggestions and comments.

AUTHOR OF THIS REVISION: William A. Rockstroh has an accumulated 40 years of experience in technical writing, editing, and technical services management. Mr. Rockstroh is currently associated with M & B Supply, Inc., a supplier of technical publications.

FOR MORE INFORMATION: This book is one of many books published on agricultural and industrial machinery. For more information or to request a FREE CATALOG, call 1-800-522-7448 or send your request to:

Deere & Company, John Deere Publishing
Almon TIAC Bldg., Suite 140
1300 – 19th Street, East Moline, IL 61244
or go to:
http://www.deere.com/deerecom/Farmers+and+Ranchers/Publications

 We have a
long-range interest in
good service

ISBN 0-86691-278-9

CONTENTS

1 FUELS

Introduction 1
Compression and Fuels 2
Selecting Fuel for Gasoline Engines 4
Selecting Fuel for LP-Gas Engines 9
Selecting Fuel for Diesel Engines 11
Storing Gasoline Fuels 17
Storing Diesel Fuels 20
Storing LP-Gas Fuels 23
Test Yourself 26

2 LUBRICANTS

Introduction 28
Engine Oils 28
Gear Oils 46
Differential Oils 48
Hydraulic and Transmission Fluids 49
Lubricating Greases 53
Test Yourself 55

3 COOLANTS

Introduction 57
Parts of Liquid Cooling System 57
Development of Liquid Cooling 58
Water as a Coolant... 58
Antifreezes 58
Coolant Filter and Conditioner 61
Maintenance of Cooling System 63
Summary 68
Test Yourself 68

4 *FILTERS*

Introduction 69
Air Cleaners 69
Fuel Filters 71
Engine Oil Filters and Filtration Systems 72
Transmission and Hydraulic System Filters 74
Coolant Filter and Conditioner 76
Test Yourself 76

APPENDIX

Answers to Test Yourself A-1
Measurement Conversion Chart A-3

FUELS

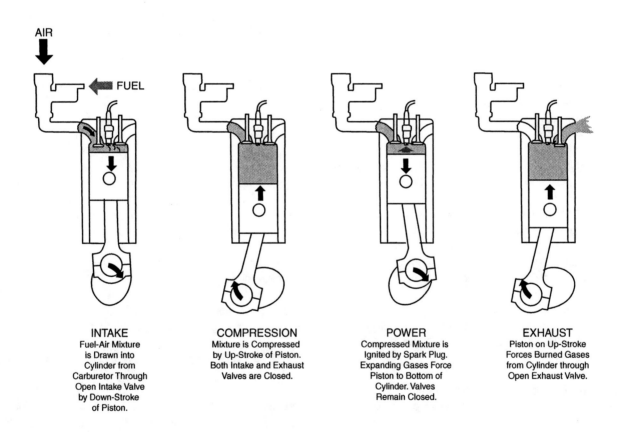

INTAKE	COMPRESSION	POWER	EXHAUST
Fuel-Air Mixture is Drawn into Cylinder from Carburetor Through Open Intake Valve by Down-Stroke of Piston.	Mixture is Compressed by Up-Stroke of Piston. Both Intake and Exhaust Valves are Closed.	Compressed Mixture is Ignited by Spark Plug. Expanding Gases Force Piston to Bottom of Cylinder. Valves Remain Closed.	Piston on Up-Stroke Forces Burned Gases from Cylinder through Open Exhaust Valve.

Fig. 1 – How a Four-Stroke Cycle Engine Works - Spark-Ignition Shown

INTRODUCTION

If you purchase a new machine today, you will find that the engine is designed for use with a certain fuel. For example, if it is a tractor equipped with a diesel engine, you have no choice but to use diesel fuel. It will not operate on gasoline or LP-Gas.

This situation has not always been true. Two-fuel engines that started on gasoline and operated on kerosene or distillate were quite popular at one time. Some are still in use but they are not as efficient as engines designed for use with one fuel.

FUELS, LUBRICANTS, COOLANTS AND FILTERS

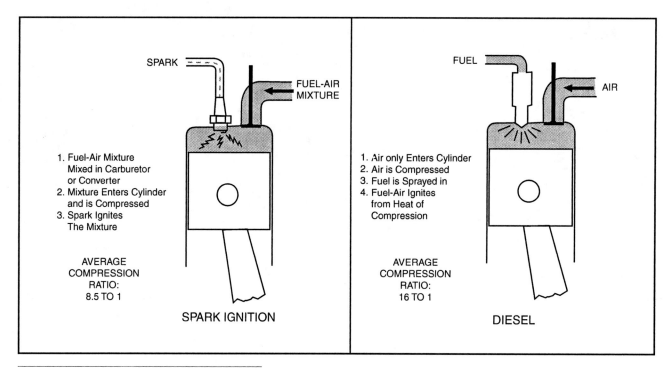

Fig. 2 – Spark-Ignition and Diesel Engines Compared

From this it might appear that there is no fuel selection problem once you have selected your engine. But the quality of each fuel you use is directly related to the maintenance required and to the performance you can expect.

Then, too, the quality of fuels and their ingredients keep changing. Various terms are used to describe these changes, such as: "octane rating," "sulfur content," "cetane rating" and "volatility." You need to know what these terms mean in order to understand how these changes apply to your engine. An example is the gradual increase in the octane rating of regular gasoline over a period of years. Is it of any value to you as a user of a gasoline engine?

In addition, spark-ignition engines require fuels with certain qualities, while diesel engines require other fuel qualities. This is because of the difference in the way the fuel is ignited in each type.

If you are not familiar with the two types of engines and how they work, study Figures 1 and 2. They show the operating principles of engines now being used on farm and industrial machines.

The important points you need to know about selection of each fuel are discussed under these headings:

1. Selecting fuel for gasoline engines

2. Selecting fuel for LP-Gas engines

3. Selecting fuel for diesel engines

COMPRESSION AND FUELS

In discussing each of these fuels, the term "compression ratio" is used frequently. It is important that you understand its meaning.

Compression ratio is the relation between the total volume inside the engine cylinder when the piston is at its greatest distance from the cylinder head (Fig. 3), compared to the volume when the piston has traveled closest to the cylinder head.

Diesel engines ignite fuel by heat resulting from a compression ratio of 16 to 1.

With lighter fuels such as regular gasoline, the compression ratio is about 8 to 1 and as high as 9 to 1.

LP-Gas, a still lighter fuel, can have a compression ratio of 9 to 1 or higher.

The higher the compression ratio, the more the fuel-air mixture is compressed and the higher the pressure inside the cylinder before the fuel burns. If the fuel burns properly, higher compression greatly increases the power output of the engine because more of the fuel energy is developed into useful power.

However, each fuel has its limit on how much it can be compressed and still burn properly when ignited in a spark-ignition engine.

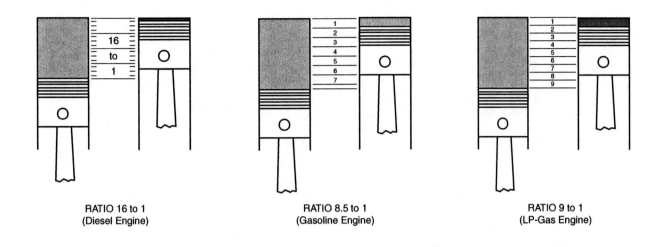

RATIO 16 to 1
(Diesel Engine)

RATIO 8.5 to 1
(Gasoline Engine)

RATIO 9 to 1
(LP-Gas Engine)

Fig. 3 – Fuels Must be Matched with Engines of Proper Compression Ratios

1. Spark Begins Fuel-Air
   Mixture Burning.

2. Flame Advances
   Smoothly, Compressing
   and Heating End-Charge.

3. End-Charge Suddenly
   Ignites with Violence,
   Producing a Knock.

Fig. 4 – How Combustion Knock Occurs in a Gasoline Engine

For example, kerosene burns evenly in an engine with a compression ratio of 4 to 1 and supplies a smooth flow of power. But if it is used in an engine with a compression ratio of 7.5 to 1, it will burn unevenly and cause the engine to "knock" (Fig. 4). During knock, fuel ignites next to the spark plug and builds up tremendous pressure on the unburned fuel as shown. This pressure then causes spontaneous combustion of the unburned fuel, causing a small explosion or knock.

Fuel *knock* is a serious problem because it may damage valves, pistons (Fig. 5) and bearings. It also results in a loss of power.

If regular gasoline is used in place of kerosene, it will burn satisfactorily with either the 4 to 1 compression ratio or the 7 to 1 compression ratio. However, it will not burn satisfactorily in an engine with a 9 to 1 compression ratio.

Fig. 5 – Combustion Knock Damaged this Piston and Broke the Second Ring. Top Ring is Stuck and Broken.

COMPRESSION RATIOS AND OCTANE RATINGS
OF SPARK-IGNITION FUELS

Fuel	Engine Compression Ratio (approx.)	Approx. Octane No.
Gasoline (Regular)	7.0–9.0 to 1	82–92
Gasoline (Premium)	9.0–10.0 to 1	92–100
LP-Gas		
Butane	8.0 to 1	100–110
Propane	9.0 to 1	110–120

Combining the octane ratings of the Motor Method and the Research Method then dividing by 2 derives the octane numbers shown in the previous chart. This gives you the average between the two. These numbers are typically found at the pump.

The table also gives the compression ratios and octane ratings expected with modern fuels. With diesel fuels, *cetane* ratings are a prime factor in starting, this will be covered later.

SELECTING FUEL FOR GASOLINE ENGINES

Gasoline as a fuel is still widely used for farm and industrial machines. But in recent years diesel engines have become more prominent in these applications. We will discuss diesel fuels later.

A factor that helped make gasoline a popular fuel was the raising of the "octane rating." When oil companies raised the octane, engines with higher compression ratios were developed to take advantage of the efficiency made possible by using the higher-octane fuels.

Almost all new, one-fuel gasoline engines are designed to operate on regular-grade/lead-free gasoline. If you purchase regular-grade/lead-free gasoline from a reliable dealer, you are almost certain to have the grade and quality of fuel needed for your engine. But it is important that you understand the principle qualities that make the fuel satisfactory as compared to low-grade (fleet) gasoline or even as compared to higher grades of gasoline.

The important qualities are:

1. **Proper octane rating**

2. **Easy starting**

3. **High oxidation stability and freedom from gum**

4. **Freedom from foreign matter, including water and dirt**

5. **Use of fuel additives**

PROPER OCTANE RATING

The octane number rating shows your assurance of getting a fuel with adequate antiknock. Your operator manual tells you what the minimum octane rating should be for your engine. Compare it with the octane rating of the fuel supplied by your dealer. He should know its octane number, but if he doesn't, he can easily find out.

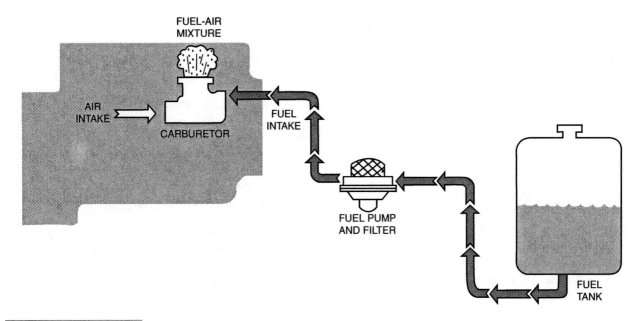

Fig. 6 – Gasoline Fuel System

If you don't understand the meaning of "octane number rating," the following discussion will help.

The **octane rating** is a method of comparing the antiknock qualities of fuels used in spark-ignition engines with standard test fuels. The ASTM (American Society for Testing Materials) has established it. The lowest number on the octane scale is 0.

Fuels near that end of the scale have a bad tendency to knock; kerosene is one of them. Fuels with the least tendency to knock have higher octane numbers, some even higher than 100. LP-Gas and super-premium gasoline are examples of these.

You may see *two octane ratings* mentioned in your operator manual. For example, one manual states: "The gasoline should have a minimum octane number rating of 85 (Motor Method) or 93 (Research Method)." This is confusing unless you understand that, while both methods use much the same test procedures and equipment, there are certain differences such as: engine speeds, spark advance and intake air temperatures. For example: With the Research Method the test engine runs 600 rpm (revolutions per minute); with the Motor Method the test engine runs 900 rpm. This results in a difference in the octane number ratings between the two test methods.

For the same fuel, the Research Method octane number is always higher than the Motor Method octane number, but not by any fixed amount. The spread may vary from narrow to wide, depending on the makeup of the fuel.

If you are given the octane rating of gasoline and only one number is given with no mention of the Method, you can assume it is the *Research Method* octane number. But, gasoline pumps are often marked with the average of the Research Method and Motor Method for consumers, (M+R / 2). You would be wise to look at the octane ratings on the gas pumps to determine which is highest and lowest.

The names "premium," "regular" and "low-grade" are rough comparative measures of octane ratings. Most engines on farm and industrial machines use regular grade. Most manufacturers design their engines to use *regular-grade* gasoline. But, regular-grade fuel has increased in octane rating. And, manufacturers have increased compression ratios in their engines to get the higher efficiencies made possible by higher-octane fuels.

You can use premium-grade gasoline, but there is usually no advantage since most engines are not designed (do not have a high enough compression ratio) to benefit from the higher octane rating, and premium costs more per gallon.

Low-grade (fleet) fuel is rarely used anymore, and it is not a good choice, because it develops less power and causes engine knock on hard pulls.

There are several ways of improving the octane rating of gasoline. For years the common way was to add "tetraethyl lead" (TEL). But environmental concerns have dictated the use of other additives. Non-leaded gasoline uses hydrocarbon fractions of high octane number to get the same results.

Gasoline containing "tetraethyl lead" is now less available. Engines designed to run on leaded fuel may experience valve recession and increased ring wear if run on unleaded gasoline.

EASY STARTING

Provision for easy starting is "built in" by both the gasoline and engine manufacturers.

The gasoline property, which is most important in engine starting and performance, is **volatility**. The volatility of a gasoline affects engine operation in a number of ways. If it is *too low*, insufficient vapor can affect starting. On the other hand, a gasoline with *too high* a volatility is apt to cause carburetor icing and also vapor lock under adverse atmospheric conditions. A balance must be struck between these extremes. The volatility is controlled and measured by a laboratory distillation test and by a Reid vapor pressure test.

In *summer*, oil companies blend their gasoline so "volatility" – its tendency to evaporate – is low. With the higher summer temperatures, your engine will start without the gasoline having high volatility. In *winter*, your engine will be slow to start unless the gasoline vaporizes readily; so oil companies blend gasoline for higher volatility.

If you have held over a supply of gasoline from summer to winter, you may experience hard starting. The only controls you have over volatility are to buy small quantities of gasoline and store them properly so as to keep down evaporation. This is discussed later in this chapter under "Storing Gasoline Fuels".

Storage Stability

The tendency of motor gasoline to form gum in storage is an indication of its oxidation stability.

Oxidation stability is tested by an ASTM method, which measures the stability of gasoline under accelerated oxidation conditions.

In this test, a special container equipped with pressure gauge is partially filled with gasoline; the remaining volume of the bomb is filled with oxygen at 100 pounds per square inch (690 kilopascals) pressure. The bomb is placed in a water bath at 212°F (100°C), and the time in minutes from the start of the test until the pressure drops 2 pounds in 15 minutes is reported as the induction period. The length of the induction period is an indication of the oxidation resistance of the sample since the pressure drop is indicative of the start of the oxidation of the unstable components in the sample. Most modern gasoline has induction periods that exceed 240 minutes.

This method was developed when unstable, thermally-cracked gasoline stocks were widely used. Its usefulness has decreased with the trend to gasoline stocks of greater inherent stability.

Antioxidant additives that minimize gum formation and lead antiknock decomposition also adequately stabilize most of today's gasoline.

FREEDOM FROM FOREIGN MATTER

Freedom from dirt and moisture is mostly a matter of how gasoline is handled and stored. Most distributors of petroleum products are well equipped and careful in handling fuels to avoid dirt. The dirt and moisture problem usually develops during handling and storage on the farm or job site. This too is discussed later in this chapter.

GASOLINE ADDITIVES AND THEIR FUNCTIONS

Additives have played an important role in gasoline since fuel containing tetraethyl lead was first marketed in 1923.

Additives have become essential ingredients of modern gasoline, except for tetraethyl lead. Widespread use of most additives has occurred during the past two decades.

Additives are used to raise octane number and to combat surface ignition, spark plug fouling, gum formation, rust, carburetor icing, and deposits in the intake system, and intake valve sticking.

In most cases, a chemical compound satisfies one of these functions. In some cases, however, an additive may perform more than one job. For example, a carburetor detergent may serve as an anti-icing agent and a corrosion inhibitor.

1. Antiknock Additives

Antiknock compounds are used to combat the tendency of a gasoline to knock in a spark-ignition engine. Ideally, gasoline should burn smoothly and evenly in the combustion chambers of an engine.

However, gasoline is composed of hundreds of different hydrocarbons, many of which may react rapidly and very violently when compressed and heated in the presence of air. After ignition, the spreading flame front in an engine's combustion chamber further compresses and heats the fuel-air mixture ahead of it. Under these conditions, some hydrocarbons in the unburned end gas may undergo chemical reactions prior to normal combustion. The products of these reactions may then self ignite. When this happens, the end gas burns very rapidly – at a rate of 5 to 25 times that of normal combustion – causing high-frequency shock waves that produce the sharp metallic noise called knock.

Besides producing an objectionable sound, a "knocking" engine may give less power and poorer fuel economy. Severe knock also tends to increase piston-ring wear and to cause overheating of valves, spark plugs, and pistons, shortening their service life and promoting destructive pre-ignition.

Factors Affecting Knock

The presence or absence of knock in an engine is determined by two factors – antiknock quality (octane number) of the gasoline and the octane number requirement of the engine.

Higher gasoline octane numbers (more resistance to knock) can be achieved in two ways:

   a) *By refining processes, such as catalytic reforming, which convert low-octane hydrocarbons to high-octane hydrocarbons.*

   b) *By using antiknock additives.*

Selection of the route to be used is dictated by cost and usually involves a combination of processing and antiknock additives.

The octane number requirement of an engine depends on how its design and operating conditions affect the temperature and pressure of the end gas in the engine.

An engine's tendency to knock varies with the compression ratio, spark advance, manifold vacuum, engine speed, combustion chamber design, air-fuel ratio, altitude, and atmospheric conditions.

TETRAETHYL LEAD (TEL)

For many years, TEL was used as a highly effective antiknock agent for the majority of gasoline blends. The precise mechanism by which TEL controls knock is not known despite extensive research.

Because of ecological and health concerns, TEL is scarcely used anymore.

The new non-leaded gasoline use hydrocarbon fractions of high octane to replace the lead additives and so provide a "cleaner" fuel. The addition of alcohol will also raise the octane level.

NOTE: Some older engines that were designed to run on leaded fuel may experience piston ring wear and valve recession, when run on non-leaded gasoline.

SCAVENGERS

When a gasoline containing lead antiknock alone is burned in a spark-ignition engine, it produces nonvolatile combustion products. Therefore, commercial antiknock fluids contain scavenging agents – ethylene dibromide and/or ethylene dichloride – that transform the combustion products of the antiknock into forms that are vaporized readily from hot engine surfaces. These scavengers are included as a part of any antiknock compound containing lead.

2. Deposit Modifier Additives

Deposit modifiers combat surface ignition and spark plug fouling by altering the chemical character of combustion-chamber deposits.

SURFACE IGNITION

Surface ignition occurs when the fuel-air charge is ignited by hot spots within the combustion chamber, this includes glowing deposits.

To the operator, surface ignition usually reveals itself either as sporadic high frequency knocking called "wild ping" or as low-frequency noises, similar to those produced by worn main bearings, called "rumble".

Wild ping results when the surface-ignited flame front causes the pressure and temperature in the unburned portion of the fuel-air charge to rise much faster than in normal combustion. Consequently, the unburned fuel-air mixture is stressed far beyond its antiknock quality and knock results.

Rumble, on the other hand, is a forum on non-knocking combustion. It occurs when ignition from a number of sources produces a very rapid pressure rise and high peak pressure during the compression stroke of the engine cycle.

In extreme cases, surface ignition can heat deposits and engine parts to the point where ignition occurs progressively earlier in the cycle. Such runaway pre-ignition can quickly burn holes in pistons or exhaust valve faces.

Phosphorus compounds are widely used as deposit modifiers. These additives suppress surface ignition by raising the temperature required to initiate glowing of deposits and by reducing the rate of heat release from oxidation of the deposits.

Some refiners use a boron compound to combat surface ignition. In addition, boron promotes the antiknock action of TEL in certain types of gasoline. This results from the ability of the boron to prevent normal sulfur concentrations in gasoline from reducing the antiknock effectiveness of TEL.

SPARK PLUG FOULING

Combustion products tend to deposit on spark plug insulators (Fig. 7) when an engine is operated under light-duty low-temperature service, such as in stop-and-go driving. Subsequent acceleration raises the temperature of these deposits. Since the electrical resistance of the deposit decreases with rising temperature, the deposits may become sufficiently conductive to prevent the plugs from firing. Such misfiring usually occurs at high engine speeds or during acceleration in the middle-speed range.

Fig. 7 – Deposit Fouling of Spark Plug

To alleviate spark-plug-fouling, phosphorus compounds in gasoline fuel change the deposits to forms having much higher electrical resistance over a wider temperature range.

3. Antioxidants

Antioxidants are added to gasoline to prevent gum formation during the normal life of the finished fuel.

Gum forms in gasoline when the unstable hydrocarbons combine with oxygen (oxidize) or with each other (polymerize). The type of crude oil from which the gasoline is produced, the refining processes used, the storage temperature, the extent to which air is present, and the length of storage influences gum formation.

When gum is formed, it produces a varnish-like deposit that tends to coat and clog the fuel lines, carburetor jets, and intake manifold, and may also cause intake valves to stick.

The oxidation stability of gasoline can be improved by various refining treatments, such as caustic washing, acid washing, partial hydrogenation, and contacting with activated clays. However, such processing is usually quite expensive and tends to rob the gasoline of the unsaturated, high-octane, olefinic hydrocarbons. Therefore, refiners frequently find it more economical to use small quantities of an antioxidant to supplement or replace processing.

Antioxidants retard the oxidation and polymerization of the unstable hydrocarbons. Although the mechanism of this action is not well defined, antioxidants are believed to act as "chain-breakers" in the various oxidation and polymerization reactions.

4. Antirust Agents

Rusting and corrosion can lead to severe problems in storage tanks, lines, and the fuel systems of engines.

For example, leaks may develop in corrosion-weakened tanks or lines, or particles of rust may impede engine operation by clogging filters and carburetor jets. In addition, a rust particle lodged on the seat of a carburetor's needle valve may cause the float bowl to overflow, followed by engine stalling due to "flooding".

Rusting and corrosion of iron and bimetallic parts are promoted by small amounts of water and air dissolved in gasoline. Water can enter the fuel system of an engine by condensation in the fuel tank or by being pumped in with the gasoline.

Several types of hydrocarbon-soluble compounds are used as rust inhibitors. These include various fatty acid amines, sulfonates, alkyl phosphates, and amine phosphates. Most of them act by coating metal surfaces with a very thick protective film that keeps water from contacting the surfaces. This "surface active" property can also help to prevent carburetor icing and the build-up of carburetor deposits.

Fig. 8 – Use Only a Good Grade of Gasoline

5. Anti-Icing Agents

Ice can interfere with engine operation either by plugging fuel lines or by upsetting carburetion through ice formation in the air or fuel passages. Plugging of fuel lines stems from water present in the fuels. Freezing of water vapor, however, causes carburetor icing, from the air the engine breathes.

When a cold engine is started under certain atmospheric conditions – the most critical being from 30° to 50°F (-1° to 10°C) at relative humidity above 65% – the cooling effect of fuel vaporizing in the carburetor causes the moisture in the air to condense and freeze on the chilled carburetor surfaces. When the throttle is almost completely closed for idling, this ice tends to bridge the small gap between the throttle blade and the throttle body, cutting off the air supply and stalling the engine. Opening the throttle for restarting breaks the ice bridge but does not eliminate the possibility of further stalling before the engine and carburetor have fully warmed up.

Two general types of carburetor anti-icing additives – *freezing-point depressants* and *surface-additive agents* – are used in much gasoline to avoid the annoyance of cold-weather stalls during engine warm-up.

The freezing-point depressants serve as antifreezes, combating carburetor icing in much the same way as antifreeze provides winter protection in an engine's cooling system. These anti-icing additives, which include alcohol, glycols, and a formamide, reduce ice formation by lowering the freezing point of the water vapor in the air.

The surface-active agents provide a different kind of protection. With these agents, ice particles are allowed to form. However, the additive provides a coating that tends to prevent these particles either from sticking to each other or from building up on the metal surfaces of the carburetor. Instead, most of the ice particles pass harmlessly through the carburetor into the intake manifold. Examples of this type of additive are amides, amines, and amine or ammonium salts of phosphates.

Some freezing-point depressants and, to a lesser extent, some surface-active agents are also effective in preventing ice plugging of fuel lines.

6. Carburetor Detergents

When an engine is idling, nonvolatile fuel components, together with contaminants from exhaust and crankcase fumes drawn in through the air cleaner, tend to accumulate on the inside walls of the carburetor just below the throttle blade. By interfering with airflow and upsetting the air-fuel ratio, these deposits can lead to rough idling with frequent stalls and reduced performance and fuel economy of the engine.

Detergent additives are used to prevent deposit buildup in carburetors and to remove deposits already formed. The effectiveness of these detergents, which include amides and alkyl amine phosphates, stems from their surface-active properties.

7. Dyes

Dyes are added to gasoline to indicate the presence of antiknocks, promote sales appeal, and identify various makes or grades of gasoline. Dye is included in leaded fuel to identify its use as a motor fuel only and to warn against its misuse for heating or cleaning purposes.

Gasoline dyes are hydrocarbon-soluble organic compounds that are selected for the color they impart to the fuel.

Dye concentration depends on the intensity of color desired by the refiner to meet a color standard.

8. Alcohol

Some producers, to form a fuel blend called Gasohol, have blended alcohol with gasoline. The fuel is usable in most gasoline-fueled engines, and there may be an economic advantage to some, especially cooperatives. However, there are some aspects of the fuel you should weigh carefully if you use an alcohol/gasoline blend.

Alcohol may attack rubber and some plastic products like fuel lines, gaskets, filters, and fittings. It also corrodes some metal parts in fuel systems. There is usually a drop in power when Gasohol is burned in an engine. Alcohol raises the vapor pressure of gasoline in tanks, and lowers the amount of air needed for an explosion. This makes Gasohol more vulnerable to flashing (igniting with a spark).

SELECTING FUEL FOR LP-GAS ENGINES

LP-Gas may be:

a) **all propane**

b) **all butane**

c) **a combination of the two gases**

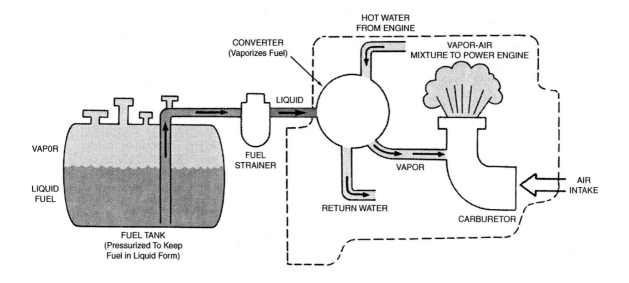

Fig. 9 – Lp-Gas Fuel System

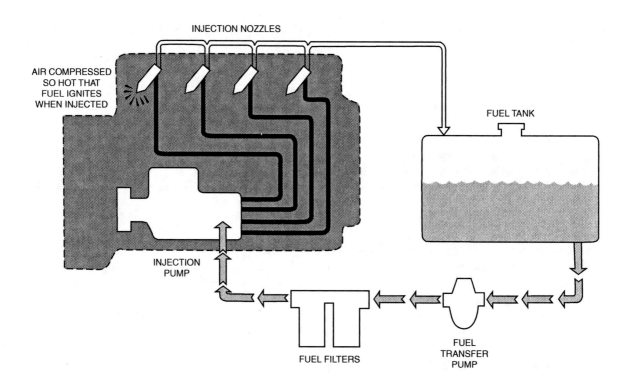

Fig. 10 – Diesel Fuel System

Today, LP-Gas is either all propane or mostly propane because of the high demand for butane in the chemical industry. Both products are gases and cannot be used through a regular gasoline tank and carburetor. The reason: They must be stored and handled in high-pressure containers to keep them in liquid form. Butane boils at approximately 33°F (1°C) while propane boils at -44°F (-42°C). When confined to a closed container, the pressure varies with the outside temperature. Butane develops 37 pounds per square inch (255 kilopascals) pressure at 100°F (38°C) while propane develops 195 pounds per square inch (1340 kilopascals) pressure at the same temperature.

An LP-Gas engine is similar to a gasoline engine, but is designed with a higher compression ratio – about 8 to 1 to as high as 10 to 1. Both gases have a high *octane rating* ranging from about 95 for butane to as high as 125 for propane. With such a high octane rating they are well fitted to high-compression engines. In fact, higher compression enables these fuels to supply power economically even though they actually contain less energy per gallon than gasoline, kerosene or "tractor fuel".

Machines equipped for LP-Gas use the vapor in the top of the fuel tank (Fig. 9) for *easy starting* because it is already vaporized.

There is very little you can do about selecting LP-Gas fuel except to deal with a reliable distributor. You are entirely dependent on him to supply fuels that are relatively *free of sulfur compounds* and *other contaminants* which may cause difficulties such as filter plugging, valve failures, etc. However, customers seem to have had very little bad experience from impurities in LP-Gas fuel.

For cold weather operation, your dealer may have a different blend of LP-Gas than for warmer weather.

SELECTING FUEL FOR DIESEL ENGINES

There is a saying that "Diesel engines will burn anything." It is true that they will burn a wide variety of fuels; in fact, powdered coal was the first fuel used for diesel engines.

From this you may gain the wrong impression – that selecting fuel for your diesel engine is simply supplying anything it will burn; but diesel fuel selection is far more critical than this. In fact, you need to know more about the qualities of diesel fuels than about the quality of gasoline.

Before studying diesel fuels selection, we must first understand how a diesel engine works. Figure 2 shows the operating principles of a 4-stroke cycle diesel engine as compared to a gasoline engine.

Note that there is no spark plug to start the fuel burning. Instead, the air is compressed until it is so hot that fuel injected into it will start burning spontaneously (Fig. 10).

The compression ratios for diesel engines are much higher than for spark-ignition engines. This extra compression of the air provides high enough temperatures (900° to 1200°F) (480° to 650°C) that the fuel is ignited by itself when sprayed into the cylinder. Note in Figure 2 that the compression ratio is 16 to 1. This compares with the compression ratios of about 8 to 1 in gasoline engines.

The average compression ratio for diesel tractors is approximately 16.3 to 1. They vary from as low as 14 to 1 to as high as 20 to 1.

THE DIESEL CYCLE

In a *spark-ignition* engine, the charge of fuel and air is taken into the cylinder as a mixture, compressed, and ignited by the spark plug.

By comparison, in the *diesel engine*, air alone is taken into the cylinder and compressed. The fuel is injected at high pressure through a nozzle into the compressed air charge, and is ignited by the heat of this charge.

During injection, it is vital that the atomized fuel particles are fully mixed with the molecules of hot compressed air so that the maximum possible number of ignition points are created throughout the charge to provide early and uniform ignition.

Depending on the compression ratio, the temperature of the intake air, and the injection timing, the temperature and pressure of the compressed air at the time the fuel enters the cylinder may be as high as 1200°F (650°C) and 900 pounds per square inch (6200 kilopascals).

The speed and power output of the diesel engine is regulated by the amount of fuel injected per stroke. At any given engine speed, the amount of air taken into the engine usually remains constant regardless of load, and there is generally sufficient air to burn the amount of fuel injected.

This means that the overall air-fuel ratio in a cylinder must always be leaner than that required to burn the fuel completely. However, since diesel engines smoke on occasion, incomplete burning will occur under certain conditions. The most likely causes of black smoke are faulty fuel injection, air restriction, turbocharger lag, or over-fueling. White smoke, which may occur during idling or other cool-engine operation, is caused by unburned or partially burned fuel.

To obtain an understanding of the major requirements for a diesel fuel, it is necessary to consider the diesel combustion process.

It would be reasonable to assume that ignition occurs the instant the fuel particles contact the high temperature air within the combustion chamber. However, this does not happen. Pressure measurements made during the compression and power strokes indicate that there is a slight delay between the start of fuel injection and the time that sufficient energy is released by burning to increase the pressure above that obtained by compression of the air alone.

The ignition delay is an important factor in diesel engine combustion. Too long a delay period at high engine loads results in a too rapid increase in pressure when the fuel starts to burn. The rate of pressure rise may become so rapid at high engine loads that knock or rough engine operation will occur. For many years this sudden increase in pressure was attributed to accumulation of fuel in the combustion chamber. However, evidence is mounting to support the theory that a long delay period allows more time for certain chemical reactions to take place in the fuel-air mixture before ignition occurs. These reactions result in products that burn very rapidly, causing excessively rapid pressure rise. With a short delay period, ignition apparently occurs before these reactions have proceeded far enough to cause too rapid burning. Also, with a cold engine and low intake air temperatures, too long a delay period produces misfiring and uneven or incomplete combustion, with consequent smoke and loss of power.

Although the ignition delay is influenced by engine operating conditions, it depends primarily on the hydrocarbon composition of the fuel and, to a lesser extent, on its volatility.

DIESEL FUEL CHARACTERISTICS

To insure the proper characteristics and maintain the uniformity of product, the refining processes used to produce diesel fuel must be closely controlled.

The American Society for Testing Materials (ASTM) has established a classification of diesel fuels for various types of diesel engine service.

The major grades covered in the ASTM "Specification for Diesel Fuel Oils" are:

Low Sulfur No. 1-D Diesel Fuel

Low Sulfur No. 2-D Diesel Fuel

No. 1-D Diesel Fuel

No. 2-D Diesel Fuel

No. 4-D Diesel Fuel

The limiting requirements by the ASTM for these grades of diesel fuels are shown in the following chart. Some of these fuel characteristics are discussed next. Remember that engine design and operation materially affect the type of fuel best suited for an engine.

Grade Low Sulfur No. 1-D

Grade Low Sulfur No. 1-D is the class of low-sulfur volatile fuel oils from kerosene to the intermediate distillates. The fuels within this grade are applicable for use in high-speed engines that require low sulfur fuel and in services involving frequent and relatively wide variations in loads and speeds, and also where abnormally low fuel temperatures are encountered.

ASTM LIMITING REQUIREMENTS FOR DIESEL FUELS												
Diesel Fuel Grade	Flash Point		Cloud Point	Water and Sediment % Vol.	Carbon Residue on 10% Residue %	Ash % Mass	Distillation Temperature, 90% Volume Recovered		Kinematic Viscosity at 40°C Centistokes (mm²/s)		Sulfur % Wt.	Copper Strip Corrosion	Cetane Number
	Min	Max	Max	Max	Max	Max	Min	Max	Min	Max	Max	Max	Min
Low Sulfur No.1-D	100°F (38°C)	*	0.05	0.15	0.01	——	550°F (288°C)	1.3	2.4	0.05	No, 3	40
Low Sulfur No.2-D	125°F (52°C)	*	0.05	0.35	0.01	540°F (282°C)	640°F (338°C)	1.9	4.1	0.05	No. 3	40
No.1-D	100°F (38°C)	*	0.05	0.15	0.01	——	550°F (288°C)	1.3	2.4	0.50	No. 3	40
No.2-D	125°F (52°C)	*	0.05	.035	0.01	540°F (282°C)	640°F (338°C)	1.9**	4.1	0.50	No. 3	40
No.4-D	131°F (55°C)	*	0.50	——	0.10	——	——	5.5	24.0	2.00	——	30

* *For cold weather operation, the cloud point should be specified at 10°F (6°C) above the tenth percentile mimimum ambient temperature at which the engine is operated except where fuel heating is provided.*

** *When cloud point less than 10° (6°C) is specified, the minimum viscosity shall be 1.7 mm 2/s and the minimum 90% distillation temperature shall be waived.*

Grade Low Sulfur No. 2-D

Grade Low Sulfur No. 2-D includes the class of low sulfur distillate gas oils of lower volatility than Grade Low Sulfur No. 1-D. These fuels are applicable for use in high-speed engines which require low sulfur fuels and in services involving relatively high loads and uniform speeds, or in engines not requiring fuels having the higher volatility or other properties specified for Grade Low Sulfur No. 1-D.

Grade No. 1-D

Grade No. 1-D is the class of *volatile* fuel oils from kerosene to the intermediate distillates. These fuels are for use in high-speed engines, in services involving frequent and relatively wide variations in loads and speeds, and also where abnormally *low fuel temperatures* are encountered.

Grade No. 2-D

Grade No. 2-D is the class of distillate gas oils of *lower volatility*. These fuels are for use in high-speed engines, in services involving relatively high loads and uniform speeds, or in engines not requiring fuels having the higher volatility or other properties specified for Grade No. 1-D.

Grade No. 4-D

Grade No. 4-D covers the class of more viscous distillates and blends of these distillates with residual fuel oils. These fuels are applicable for use in low- and medium-speed engines involving sustained loads at substantially constant speed.

Cetane Number

The method for determining the ignition quality of diesel fuel in terms of *cetane number* is similar to that used for determining the antiknock quality of gasoline in terms of octane number.

As in the case of the octane number scale, the scale of cetane number represents blends of two pure hydrocarbon reference fuels. Cetane is a hydrocarbon with very high ignition quality, and was chosen to represent the top of the scale with a cetane number of 100. The hydrocarbon called alphamethylnaphthalene has very low ignition quality, and was chosen to represent the bottom of the scale with a cetane number of zero. Blends of these two hydrocarbons represent intermediate ignition qualities, and their cetane number is the percentage of cetane in the blend. For example, a reference fuel blend containing 30 percent cetane and 70 percent alphamethylnaphthalene is assigned a cetane number of 30.

The engine used in cetane number determinations is a standardized (ASTM) single-cylinder, variable-compression-ratio engine with special loading and accessory equipment.

Under test, the compression ratio of the test engine is varied until combustion starts at top dead center. With the start of fuel injection timed at 13°BTDC and with combustion timed to start at top dead center, an ignition delay period of 13° (2.4 milliseconds at 900 rpm) is produced.

This procedure is then repeated using reference fuel blends until the unknown fuel is bracketed between two reference blends differing by not more than 5 cetane numbers. Then, knowing the cetane numbers of the bracketing blends and the compression ratios required by the fixed delay period for both the reference blends and the sample fuel, the cetane number of the sample can be obtained by calculation.

The cetane number of a diesel fuel depends primarily on its hydrocarbon composition. In general, the aromatic hydrocarbons are low in cetane number, the paraffin has high cetane numbers, and the naphthenes fall somewhere in between. Thus, it is apparent that the base stocks and refining processes used in making diesel fuels are all-important in determining ignition quality. The refiner is constantly faced with the problem of blending stocks to achieve adequate ignition quality without sacrificing other necessary characteristics, such as pour point and volatility.

The ignition quality of a diesel fuel can also be improved by the use of additives, such as amyl nitrate. The relationship between the cetane number of a diesel fuel and the performance of a diesel engine should not be confused with the relationship between the octane number of gasoline and the performance of a spark-ignition engine.

In gasoline engines, raising the octane number improves potential engine performance via increased compression ratio or supercharging, and the octane number requirement is usually fixed by the full-load demand of the engine.

In *diesel engines*, the requirements for good ignition quality during starting and light load operation at low temperatures establish the desirable cetane number.

In general, *high-cetane* fuels permit an engine to be started at lower air temperatures, provide faster engine warm-up without misfiring or white smoke, reduce the rate of formation of varnish and carbon deposits, and eliminate combustion roughness or diesel knock.

However, too *high* cetane numbers may lead to incomplete combustion and exhaust smoke if the ignition delay period is too short to allow proper mixing of the fuel and air within the combustion space.

Knock in Diesel Engines

In a *spark-ignition engine* where the fuel-air charge is mixed and then compressed before ignition takes place, *knock is caused by the fuel burning TOO RAPIDLY*. To control the rate of burning, tetraethyl lead is added.

In *diesel engines, knock is due to the fuel igniting TOO SLOWLY* (Fig. 11). It should start to burn almost as soon as it is injected (Fig. 11-A). If there is much delay, a fuel build-up results, which burns with explosive force (Fig. 11-B) and causes knocking.

INJECTOR

(a) PROPER BURNING

(Fuel Charge Ignites Early and Burns Evenly
to Overcome Knocking)

(b) POOR BURNING

(Ignition of Fuel Charge is Delayed
Followed by a Small Explosion)

Fig. 11 – In Diesel Engines Knock is Due to Fuel Igniting too Slowly

Good diesel fuel provides for *early* spontaneous combustion. By contrast, good gasoline fuels avoid spontaneous burning.

The difference between octane numbers (gasoline) and cetane numbers (diesel fuel) are shown in Fig. 12.

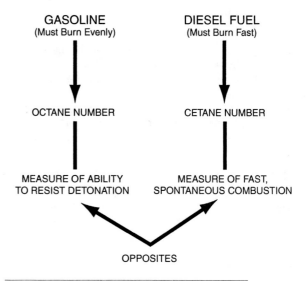

GASOLINE (Must Burn Evenly)	DIESEL FUEL (Must Burn Fast)
OCTANE NUMBER	CETANE NUMBER
MEASURE OF ABILITY TO RESIST DETONATION	MEASURE OF FAST, SPONTANEOUS COMBUSTION

OPPOSITES

Fig. 12 – Octane and Cetane Number are Opposites

Volatility

The distillation characteristics of a diesel fuel are essential for good combustion in the diesel engine. The characteristics are achieved by a careful balancing of the light and heavy petroleum fractions during the refining process. The components of the blend, which boil at the highest temperatures, have higher heating values than do the lighter fractions.

Although a lot of *heavy* fractions in the final product may improve good fuel economy, it can be harmful due to deposit formation within the engine.

Too many *light* fractions may provide easier engine starting and more complete combustion under a variety of engine conditions. However, the light ends are generally low in ignition quality. Moreover, they do not release as much energy per gallon as do the heavier fractions.

Volatility characteristics also influence the amount and kind of exhaust smoke and odor. The temperatures at which 10%, 50%, and 90% of the fuel is evaporated during a distillation test are important control points in achieving best volatility balance.

Pour Point

Diesel fuel must be able to flow at the lowest expected atmospheric temperatures. The lowest temperature at which the fuel ceases to flow is known as its *pour point*.

As fuel temperature decreases toward the pour point, the fuel becomes sluggish and harder to pump through the fuel supply lines, fuel filters, and injection system.

The pour-point temperature of a fuel is closely related to the molecular structure of its component hydrocarbons. For example, the naphthenes tend to have low pour points and relatively low cetane numbers, whereas the paraffins tend to have high pour points and relatively high cetane numbers.

Since low pour points often can be obtained only at the expense of lower cetane number or higher volatility, the pour-point specification should not be any lower than necessary.

Cloud Point

Diesel fuel becomes cloudy and forms wax crystals and other solid substances at some temperature above the pour point. The temperature at which clouding begins is called the *cloud point*.

Since the wax crystals clog fuel filters and supply lines, and since this occurs at temperatures above the pour point, the cloud point may be even more important in a fuel specification than the pour point. Like the pour point, the cloud point depends on the hydrocarbon composition of the fuel.

Although cloud points seem to occur from 8° - 10°F (5° - 6°C) above the pour points, cloud point as high as 15° or even 20°F (8° - 10°C) above the pour points are not uncommon.

Viscosity

Diesel engine injection pumps perform most effectively when the fuel has the proper "body" or *viscosity*. Lower viscosity may require more frequent maintenance of injection systems parts. Viscosity also has some influence on the atomization of the fuel when it is injected.

If the viscosity is too high, excessively high pressures can occur in the injection system. Therefore, the proper viscosity or resistance to flow of diesel fuel becomes a prime requirement.

Density and Gravity

The *gravity* of diesel fuel is an index of its *density* or weight per unit volume (kg/L at 15°C). The denser the fuel, the higher is its heat content. Since fuel is purchased on a volume basis, gravity is used in purchasing specifications and sometimes in delivery inspections. Gravity may also be measured in "degrees API". The higher the API gravity of the fuel, the lower is its density or specific gravity. In general, the refiner's choice of blending stocks and refining processes to achieve the desired volatility characteristics and cetane number determines what the gravity will be.

Flash Point

The *flash point* is the temperature to which the fuel must be heated to create a sufficient mixture of fuel vapor and air above the surface of the liquid so that ignition will occur when the mixture is exposed to an open flame. Various states and insurance companies have mandatory requirements for flash point based on fire hazard, which must be taken into account when writing diesel-fuel specifications.

Carbon Residue

The tendency of a diesel fuel to form carbon deposits in an engine may be roughly approximated by determining the carbon residue of the fuel. *Carbon residue* is the amount of material left after evaporation and chemical decomposition of the fuel has taken place at an elevated temperature for a specified period of time. High carbon residue values indicate the possibility of increased combustion chamber deposits and exhaust smoke.

Although carbon-residue tests are fairly effective in predicting deposit formation of base fuels, they are not good predictors when the fuels contain ignition improvers. Any specification for carbon residue should be stated for the base fuel.

Sulfur

Diesel fuel contains varying amounts of *sulfur*, depending on the crude oil source, refining processes, and grade. Sulfur tends to be more prevalent in the higher boiling range fractions. Sulfur content should not exceed 0.5%. Sulfur content with less than 0.05% is preferred. If diesel fuel with greater than 0.5% sulfur content is used, reduce the service interval for engine oil and filter by 50%. DO NOT use diesel fuel with sulfur content greater than 1.0%.

High sulfur content can become a problem in diesel engine operation at low temperatures and during intermittent engine operation. Under these conditions, where more moisture condensation can take place, using fuels containing excessive amounts of sulfur may cause cold corrosion and increased engine wear.

Ash

Small amounts of non-burnable material are found in diesel fuel in the form of soluble metallic soaps and abrasive solids. Since diesel engine injectors are precision-made units of extremely close fits and tolerances, they are sensitive to any abrasive material in the fuel.

The ASTM method used to determine the amount of these materials in a diesel fuel measures the *ash* content of the fuel. Burning a small sample of the fuel in a weighed container until all of the carbon matter has disappeared, as shown by the container and residue reaching a constant weight does this. The weight of the unburnable residue divided by the weight of the original fuel sample and multiplied by 100 is reported as the ash content of fuel (percent by weight).

Recommended Diesel Fuels for Farm and Industrial Machines

Below is a typical recommendation from a tractor operator's manual:

Type of Engine Service	Ambient Air Temperature	Diesel Fuel Grade No.
Light load, low speed, considerable idling	Above 80°F (27°C)	2-D
	Below 80°F (27°C)	1-D
Intermediate and heavy load, high speed, minimum of idling	Above 40°F (4°C)	2-D
	Below 40°F (4°C)	1-D
At altitudes above 5,000 ft.	All	1-D

Diesel Fuel vs. Furnace Fuel

The question is often asked if *diesel fuel* and *furnace fuel* are the same. It is possible for your fuel supplier to have one fuel that he supplies for both diesel-engine use and heating purposes. Where there is one dominant use, such as home heating compared to limited diesel-fuel use, the dealer may not be justified in stocking a separate diesel fuel. So he purchases fuel from his supplier to meet the specifications for both fuels. The refiner can meet this demand because the specifications for both fuels are broad enough that they overlap.

If a dealer is supplying big quantities of fuel for both heating and diesel engines, he will normally stock two fuels. In this case the furnace oils usually contain more of the heavy cracked distillates, which are quite satisfactory for furnaces but are not suitable for diesel engines (especially with light intermittent loads) and they do not generally meet ASTM standards for diesel engines.

Cold Weather Starting

The higher the cetane number, the easier your machine will start on diesel fuel. Ether has a cetane rating of 85 to 96 and is highly volatile. For those reasons, ether fluid is used for starting some diesel engines in cold weather (Fig. 13).

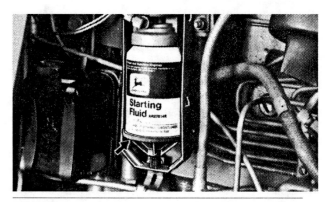

Fig. 13 – Starting a Diesel Engine in Cold Weather Using Ether Starting Fluid

IMPORTANT: Do not use thermal starting aids in the cylinder head or intake manifold when ether is used to start a cold diesel engine. Glow plugs in the turbulence chamber, intake manifold heating cold and other thermal aids can cause ether to ignite prematurely, and cause damage to the engine. If you have an ether can mounted in the fuel system, spray ether into the engine *only* while cranking with the starter. Too much ether can cause an uncontrolled explosion in the cylinder chambers and damage the engine. Ether must never come into contact with a thermal starting aid.

STORING FUELS

Your problem of storing fuels will depend on the type of fuel you are using. We have divided this discussion into three parts:

1. Storing gasoline fuels

2. Storing diesel fuels

3. Storing LP-Gas fuels

Each state has its own laws regarding the handling, storage and use of fuels. It is important that you become acquainted with them for your own safety and for insurance purposes.

In general, state laws are based on the standards and codes established by the National Fire Protection Association. It is their recommendations that are followed in this discussion.

Proper storing and handling of fuels can affect your safety, how easily your machine starts, and how much maintenance your fuel system requires.

A dependable fuel supplier knows the importance of delivering fuel to your farm or job site in top-quality condition. He is usually well trained and equipped to do the job. But, regardless of the care he uses, improper storage and negligence on your part can offset it.

Consider these four factors in the storage of fuels:

• **Protection of fuel quality**

• **Safety**

• **Convenience**

• **Cost**

The importance of these factors varies with the type of fuel you wish to store. For example, maintaining fuel quality is a serious problem with gasoline and diesel fuel, but it is not a problem with LP-Gas. However, with LP-Gas, safety is a bigger factor.

STORING GASOLINE FUELS

MAINTAINING FUEL QUALITY

You must control three conditions if you are to maintain satisfactory quality in your gasoline:

a) *Control loss through evaporation.*
b) *Avoid gum deposits.*
c) *Protect from water and dirt.*

Controlling Evaporation Losses

Evaporation losses are important both because of fuel loss and because of increased difficulties in starting your engine.

As mentioned earlier, your fuel supplier provides different gasoline blends depending on the season of the year.

In *winter*, the blend consists of more volatile fuels that provide for easier starting of your engine. In *summer*, when starting an engine is no problem, gasoline contains less of the lighter fuels; this keeps down evaporation losses.

But, if you hold your summer-blend fuel into the winter months, your machine will be hard to start. If you hold your winter-blend fuel for several months, the lighter-weight portions evaporate and will not be satisfactory for cold-weather starting. A Purdue University report states that "A sample of gasoline in storage six months in an aboveground vented tank.... would not be considered serviceable."

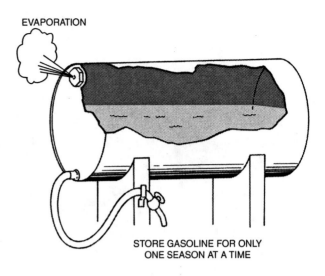

EVAPORATION

STORE GASOLINE FOR ONLY ONE SEASON AT A TIME

Fig. 14 – Lighter Fuels in Gasoline Blend for Easy Starting are First to Evaporate During Storage

Evaporation losses are sizeable from an *aboveground tank* unless you make some provision for shading it. Note in Fig. 15 that there are substantial losses of fuel from evaporation if the storage tank is exposed to direct sunlight and if the vapors are free to pass through an open vent.

These illustrations are based on studies where it was found that evaporation is especially rapid at temperatures above 90°F (32°C). This accounts for the fact that a tank painted red (Fig. 15-A), which heats rapidly when exposed to sunlight, has a greater loss of fuel than a tank painted a reflective color such as white or aluminum (Fig. 15-B). However, a tank may be almost any color if it is shaded from direct sunshine as in Fig. 15-C.

It was also shown that winter evaporation losses are about the same as those in summer. The reason: winter-blend fuel, with its larger amount of lighter-weight gasoline, evaporates faster (Fig. 14). This makes your machine start more easily on cold days. But when the sun shines, and the tank warms, your fuel losses are about equal to those in summer.

A good shade tree might provide enough protection for a fuel tank in summer but according to the report "...it is possible that some of the loss due to evaporation is the result of wind motion around open vents. This may siphon gasoline vapors from the tank. Sheet metal shades may serve to block off the wind and reduce gasoline losses in this manner."

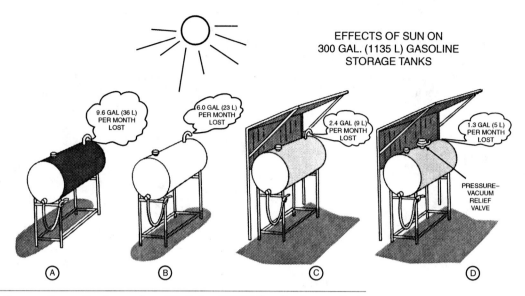

Fig. 15 – Summer Evaporation Losses from 300-Gallon (1135 L) Gasoline Storage Tanks

Evaporation losses can be further reduced by use of a *pressure-vacuum release vent.* (Fig.15-D)

If you are not familiar with the operation of a pressure-vacuum valve, note Fig. 16. It can be used in place of the standard vented cap – the type in use on most tanks. Or if your tank has a separate vent (Fig. 15-D), it can be used in place of the vent. If you should use the pressure-vacuum valve as a cap replacement and your tank has a separate vent, plug the vent. The tank must be airtight for the valve to work.

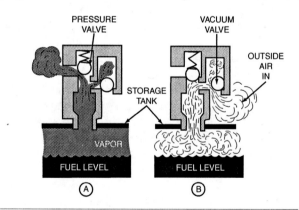

Fig. 16 – Pressure-Vacuum Relief Valve to Vent Gasoline Storage Tanks

There are thousands of pressure-vacuum relief valves in use, but before installing one, check with your state fire marshal. In some states it is against the law to use them. The reason: the pressure-vacuum valve is considered as being a type of obstruction and the National Fire Protection Code states: "Vent pipes two inches or less in nominal inside diameter shall not be obstructed by devices that will reduce their capacity and thus cause excessive back pressure." Some claim the valves require occasional cleaning which most customers tend to overlook.

Some manufacturers, in an effort to meet these objections, have increased the size of the vent area and decreased the pressure required to operate the vacuum valve to as low as one ounce to provide a minimum of obstruction. This may help the situation, but it will probably take a few more years of experience with field installations before there is a general agreement regarding its use.

⚠ **CAUTION: Don't try sealing the vent opening in your present tank to save the cost of a pressure-vacuum valve. Under hot conditions the tank pressure can increase to a point where it might burst your tank, or if the tank is hot when sealed and then cools, a vacuum tends to develop which may cause it to collapse.**

The *pressure portion* of the *valve* allows only three pounds per square inch (21 kilopascals) of pressure to develop in the tank. Above that, the pressure valve opens and allows enough vapors to escape to maintain the pressure at three pounds (Fig. 16-A). When the tank cools, the *vacuum valve* is forced open by atmospheric pressure so that air may enter as shown in Fig. 16-B. From one ounce to three pounds (0.4 to 21 kilopascals) of pressure may be required to open it, depending on the vacuum-valve setting.

If you use an *underground tank* (Fig. 17), the temperature of the stored fuel remains low enough all through the year so that the evaporation losses are small. Further provision for reducing evaporation losses is not necessary.

Avoiding Gum Deposits

The second point to consider is that gasoline will oxidize and form *gum deposits* if kept for long periods.

Refiners of gasoline add an inhibitor that will protect the fuel for six months to a year under normal storage conditions, but the time is greatly reduced if the gasoline is exposed to sunlight and to high storage temperatures. There are also certain metals, like copper, which cause gum to form faster.

Potential gum deposits can be reduced. Avoid storing more gasoline than you can use conveniently in a period of about 30 days. This too is less of a problem with underground tanks since the fuel remains cooler.

Protecting Against Water and Dirt

The same precautions you take in keeping down evaporation losses also help reduce *moisture condensation* in the tank. Moisture condensation occurs more often with aboveground tanks. The more the temperature of a storage tank varies the more air it "breathes in and out". The fresh warm air that is "breathed in" may contain more moisture than it can hold when the temperature drops. This causes moisture to condense on the inside of the tank and collect at the bottom under the fuel. The water must be drained or pumped out occasionally to avoid freezing, rusting and carburetor clogging.

A hand pump is used to remove water from an underground tank (Fig. 17).

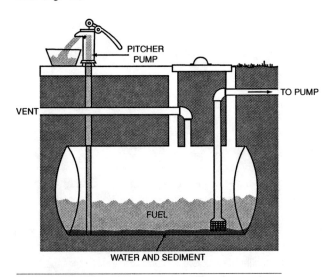

Fig. 17 – Using Hand Pump to Clean Dirt and Water from Underground Storage Tanks

There is some evidence that long storage reduces the octane number of gasoline but the reduction is not enough to concern a user.

SAFETY

 CAUTION: If you are using an aboveground tank, it should be located 40 feet (12 m) or further from the closest building as a fire safety precaution. This is the recommendation of the National Fire Protection Association. However, you may locate an underground tank so the nearest point of the shell is as close as one-foot (0.3 m) from the wall (Fig. 18).

The NFPA standard for machine fuel storage requires that aboveground tanks which feed by gravity be equipped with valves that close automatically in case of fire. In some states, law requires the valves. If your tractor catches fire, or spilled fuel ignites while you are filling the fuel tank, the fuse link melts and a spring automatically closes the valve.

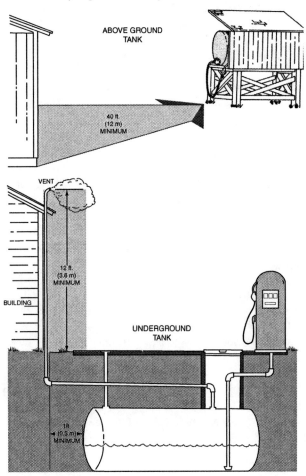

Fig. 18 – Safe Location of Gasoline Storage Tanks in Relation to Buildings

Aboveground tanks should be supported on well-built racks or on timbers or blocks, to keep them off the ground at least 6 inches (150 mm) so as to prevent corrosion of the tank.

If you install an underground storage tank, it is best to place it in a well-drained area. If it must be located where there may be high ground water levels or flooding, provide an anchor or weight it with concrete. This will keep a partially or completely empty tank from floating out of the ground.

To protect an underground tank from outer corrosion, coat it with an asphalt tar coating or some type of mastic coating recommended for that purpose. Avoid backfilling with cinders or ashes. They corrode metal that isn't well protected.

The ground should slope away from a tank or pump so vapors can drain away. Gasoline vaporizes readily. Since the vapor is 2 to 4 times heavier than air, it will collect in low places. A spark, or a match, will easily set it afire since air-gasoline mixtures of about 1.5 to 7 percent gasoline vapor will burn readily.

Gasoline-air mixture richer than about 7 percent is too rich to burn. This is the situation with a vented gasoline tank and explains why flames burn freely at the vent opening but won't follow the vent pipe down into the tank and cause an explosion.

Label your gasoline tank, particularly if there are other tanks in the vicinity containing other kinds of fuel. A **red** tank indicates it contains gasoline.

 CAUTION: If the tank is exposed to sunlight and it was painted white to reduce temperature, paint the word "GASOLINE" on it in red. Make sure the word "GASOLINE" is clearly visible when fueling in order to prevent confusion. When using an underground tank, paint the pump red or label it "GASOLINE".

CONVENIENCE

For convenience, locate your fuel storage in the vicinity of your machinery storage building. If possible, allow about 50 feet (15 m) on one side of the storage tank or pump, for maneuvering your machines and equipment. This also allows plenty of room for the tank truck to get in and out.

Convenience may help determine whether you select an aboveground tank or the underground type since the latter can be located near the outside of the building where you store your machines.

COST

An aboveground tank is much less expensive than an underground tank of the same capacity. In fact, in many localities the fuel supplier lends an aboveground storage tank to his customers as long as they purchase fuel from him. However, if you have a choice of getting either type of storage tank installed by your dealer or if you must buy a storage tank, it is well to consider the following points along with tank cost:

Advantages of Underground Tank

1. *Less fuel evaporation.*
2. *Less water condenses in the tank.*
3. *Less gum deposits form in the fuel.*
4. *Fire hazards are reduced.*
5. *The tank is hidden, avoiding a clutter around your buildings.*

Advantages of Aboveground Tank

1. *Costs less to purchase.*
2. *Not affected by ground water and limited flooding.*
3. *Is easily moved.*

The size of tank you install is the same with either type of storage since you should not install one that holds more than 30 days fuel supply.

STORING DIESEL FUELS

MAINTAINING FUEL QUALITY

Evaporation during storage (whether above or below ground) is not rapid enough with diesel fuel to be of concern. What evaporation does occur has little effect on fuel quality.

Of greatest importance in maintaining the quality of diesel fuels is that you:

a) *Keep them free of dirt and water.*
b) *Avoid gum deposits.*

Keeping Fuel Free of Dirt and Water

It is important that all fuels are kept *free of dirt and water*, but it is especially important with diesel fuel. The reason: the fuel injection system on a diesel engine is fitted with parts that are held within millionths of an inch clearance. Very fine dirt particles soon ruin the parts and cause an expensive repair job. Water; even extremely small amounts, causes corrosion, which ruins the highly polished surfaces of the injector nozzle. All operator manuals for diesels emphasize the importance of clean fuel.

Water is about the same weight as diesel fuel so it settles out very slowly. For this reason, be sure to provide some arrangement that will allow 24 hours for water and dirt to settle to the bottom of the storage tank after it has been refilled by your supplier. Use of two storage tanks is best. When one is refilled, you can use from the other to supply fuel for your machines.

If you have only one tank for storage, be sure you fill your machine fuel tank before your supplier refills the storage tank. The filling process disturbs the water and dirt particles in the bottom of the tank and mixes them with the fuel.

If you are using *drums* for storing your fuel, be sure that they are mounted rigidly in place. Any handling will remix the dirt particles and water from the bottom of the tank with the fuel.

If you are using drums as *portable tanks* for refueling in the field, be sure they are in place at least 24 hours before fuel is drawn from them.

Don't let water collect on top of your fuel storage as shown in Fig. 19-C. There are two reasons for this:

1. Water retained on the tank tends to rust the outside of the drum.

2. As fuel is drawn from the tank, water may be drawn through the air vent directly into your fuel supply.

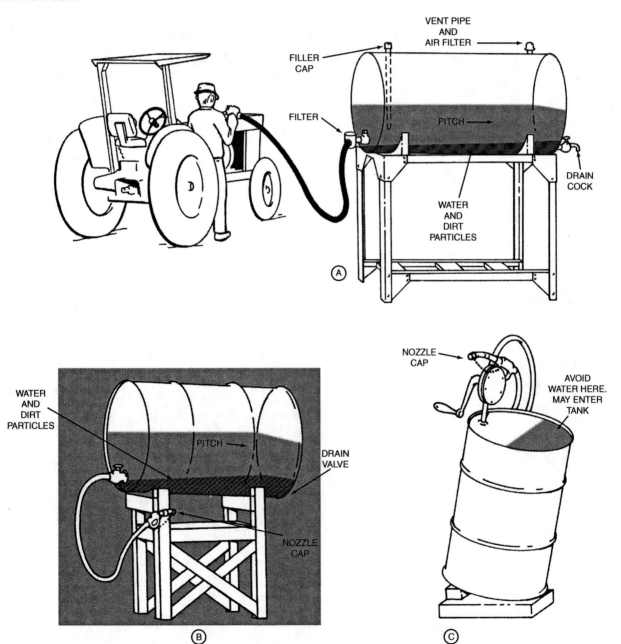

Fig. 19 – Types of Aboveground Storage Tanks for Diesel Fuel

Dirt particles may come from several different sources. Some may be present in the fuel when your supplier delivers it, but they are far more likely to come from carelessness or improper storage. Some rules for preventing dirt in the fuel supply:

1. *Don't use an open container to transfer fuel from the storage tank to the machine tank.* This greatly increases the chance for dirt to enter the fuel tank. Equip your aboveground tank with a pump and hose (Fig. 19-C) or a gravity hose (Fig. 19-A and Fig. 19-B) to transfer fuel. Then be sure to cap the end of the hose nozzle while the hose is not in use.

2. *Don't store diesel fuel in a galvanized tank.* A galvanized tank is satisfactory for gasoline, but when diesel fuel is stored in one, the fuel reacts with the galvanized finish, causing powdery particles to form. They soon clog the fuel filters on a diesel engine. Use steel tanks to avoid these troubles.

3. *Don't use a tank formerly used for gasoline storage.* Fine rust and dirt particles that settled out of gasoline and accumulated on the bottom of the tank, mix readily with diesel fuel and may remain suspended in it until drawn from the tank.

4. *Don't let the suction pipe to the fuel pump extend to the bottom of the storage tank.* This allows the pump to pick up water and sediment that has settled out of the fuel.

The end of the pipe should be 3 to 4 inches (75 to 100 mm) from the bottom. If possible, slope the tank away from the pipe or outlet valve (Fig. 19-A).

5. *Always drain the storage tank before refilling and you should clean the tank regularly.* This may allow the dirt-and-water residue to rise high enough to be drawn out with the fuel.

Draining of water and sediment from an aboveground storage tank is best accomplished by means of a valve at the bottom of the tank (Fig. 20). Don't depend on a drain plug. It is awkward to use and may result in fuel spillage and fire danger.

About twice a year (in the spring and again in the fall) thoroughly clean the tank as shown in Fig. 20. Rinse out loose sediment with clean diesel fuel. Contaminated fuel drained from the tank should be kept in a container for about 24 hours to allow dirt and water to settle. Then the clean portion of the fuel can be poured off and returned to the clean storage tank.

If you are using an underground tank, remove the water and sediment with a hand pump (Fig. 17). Twice a year is an acceptable practice.

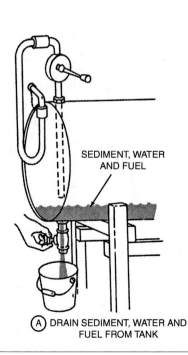

SEDIMENT, WATER AND FUEL

(A) DRAIN SEDIMENT, WATER AND FUEL FROM TANK

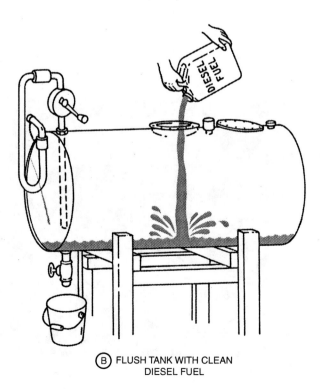

(B) FLUSH TANK WITH CLEAN DIESEL FUEL

Fig. 20 – Clean Diesel Storage Tanks Twice a Year

Avoid Gum Deposits

To keep down gum-and-varnish forming tendencies in an aboveground tank, keep it shaded from direct sunlight in the same manner as for gasoline storage (Fig. 15). This will also help to keep moisture from condensing in the tank. The fuel contains a "gum inhibitor" added by the refiners or distributor which retards the formation of gum and varnish for about three months.

SAFETY

The spacing of storage for diesel fuel is the same as for gasoline storage – about 40 feet (12 m) or more from the nearest building for aboveground storage tanks and one foot or more from the foundation wall for underground storage tanks (Fig. 18).

Many customers think that aboveground storage tanks for diesel fuel may be located next to a building. This probably comes from observing approved tank installations for oil-burning equipment installed next to a building.

But machine tanks must be filled frequently, and even though diesel fuel is much safer than gasoline, the National Fire Protection Association states that: "The occurrence of spills in handling flammable liquids is recognized as presenting the greatest potential source of the release of flammable vapors."

 CAUTION: Always check local codes for proper installation.

CONVENIENCE AND COST

When considering convenience and cost for diesel fuel storage, the same general points are involved as for gasoline storage. The exception is that evaporation is not much of a problem.

These fuels may be stored for as long as three months without seriously affecting fuel quality, while gasoline should not be stored longer than about 30 days. This may justify the use of a larger storage tank if there are times in the year when you can buy fuel at substantial savings.

STORING LP-GAS FUELS

LP-Gas must be stored in pressure-type tanks. At ordinary temperatures it changes to gas unless kept under pressure. Consequently, storage of LP-Gas is an entirely different problem from storing gasoline and diesel fuel.

There is no problem of protecting fuel quality. And there is virtually no evaporation from the pressure tank, nor does the fuel change chemically during storage. This enables you to have as large a tank as you wish and keep the fuel for as long you wish.

SAFETY

Since LP-Gas is kept under pressure and is highly flammable, rigid standards have been established by the National Board of Fire Underwriters for construction of LP-Gas storage tanks. You can tell if your tank has met Underwriters Laboratories' standards by the label on the nameplate (Fig. 21).

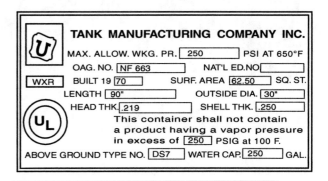

Fig. 21 – Nameplate on LP-Gas Storage Tanks has Safety Information that is Important to You

The nameplate also tells whether the tank is designed for aboveground or underground use, its working pressure in pounds per square inch, the capacity of the tank, the name of the supplier and other information you need to know.

Aboveground tanks are considered best for farm use because it is easy to find leaks (by use of a soapsuds solution) in case any develop. You should use soapsuds to check for leaks at least once each season.

 CAUTION: Don't use a flame for detecting leaks. Use a soapsuds solution to check for leaks once each season. LP-Gas has an odor additive to help you detect leaks.

With LP-Gas, as with gasoline, the storage tank must be located a safe distance from major buildings or from other fuel storage tanks. Figure 22 shows the minimum distances as provided by the standards established by the National Board of Fire Underwriters.

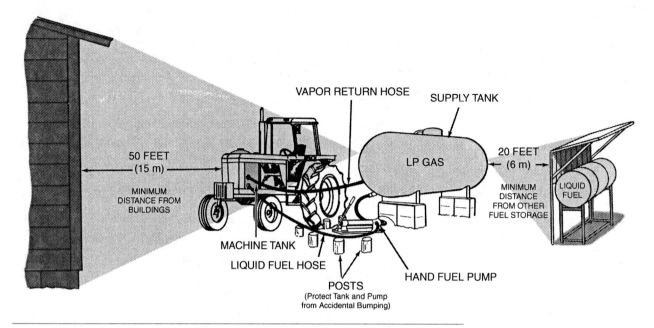

VAPOR RETURN HOSE

SUPPLY TANK

LP GAS

50 FEET
(15 m)

MINIMUM
DISTANCE FROM
BUILDINGS

20 FEET
(6 m)

MINIMUM
DISTANCE
FROM OTHER
FUEL STORAGE

LIQUID
FUEL

MACHINE TANK

LIQUID FUEL HOSE

HAND FUEL PUMP

POSTS
(Protect Tank and Pump
from Accidental Bumping)

Fig. 22 – Safe Location of LP-Gas Storage Tanks in Relation to Buildings and Other Fuel Storages

You will also need to protect from accidental damage such parts as the hose connections and pipe fittings that extend from the tank and pump, if a pump is used (Fig. 22). (If the liquid level in the storage tank is below the level of the machine tank, a pump is used to make the fuel transfer the same as with other fuels.)

Fill holes and low spots in the vicinity of the storage. Gas that leaks from the tank or fuel that escapes while filling the machine tank is heavier than air. It will drain into low spots where it will accumulate and become a fire hazard.

Don't touch LP-Gas with your bare hands as it "freezes" and can burn the skin. Wear leather gloves when handling LP-Gas containers.

CONVENIENCE AND COST

After you have figured the safety precautions; pick a location that will be easiest for refueling your machine and for refilling the storage tank. Allow about 50 feet of space on one side for maneuvering your machines and equipment and for the tank truck to refill the storage tank.

Since an LP-Gas storage tank is rather expensive, you will want to consider your cost carefully. You will need to provide:

1. For enough reserve capacity to carry through periods of heavy use without having your supplier make a special trip.

2. For taking advantage of seasonal discounts or quantity discounts, if the price quotations are low enough during some seasons of the year to justify a large tank investment.

Standard tank sizes are 250, 500 and 1000 gallons (945, 1890 and 3785 liters). *LP-Gas tanks should be filled to only 80 percent of their capacity* so the usable capacity of these tanks would be 200, 400 and 800 gallons (755, 1515 and 3030 liters), respectively. This extra space is to allow room for expansion of the liquid on hot days without developing dangerous tank pressures.

In figuring the amount of LP-Gas you are likely to use, you can estimate it as being about 25 percent more than the amount of gasoline you would use in the same machine. In arriving at tank size, don't plan on getting a tank that will require filling more frequently than every 30 days. Allow several days' reserve capacity for seasons of heaviest use.

If you are already using LP-Gas for other purposes, you have some idea as to how long your present tank capacity will take care of these needs. Determine how much this usage is on a monthly basis and add the maximum amount your machines will use. Then add at least 20 to 25 percent extra capacity to set the minimum size tank you should use.

In some areas, where LP-Gas is used for house heating, crop drying, brooding, etc., the cost per gallon is often reduced during the season of least use. You may find it pays to increase the size of tank you install to take advantage of these lower fuel prices if they are available in you area.

HANDLE FUEL SAFELY – AVOID FIRES

CAUTION: Handle fuel with care – it is highly flammable. Do not refuel the machine while smoking or when near an open flame or sparks. Always stop the engine before refueling the machine. Fill the fuel tank outdoors.

Prevent fires by keeping the machine clean of accumulated trash, grease, and debris. Always clean up spilled fuel.

Fig. 23 – Always Handle Fuel with Care

PREPARE FOR EMERGENCIES

CAUTION: Be prepared if a fire starts. Keep a first aid kit and fire extinguisher handy.

Keep emergency numbers for doctors, ambulance service, hospital, and fire department near your telephone.

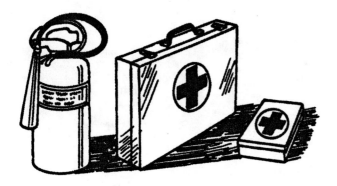

Fig. 24 – Be Prepared for a Fire

Be sure you purchase both a *liquid-fill* hose and a *vapor-return* hose for your storage tank. When you fill your machine tank with liquid; the vapor from the top of the machine tank should be fed back to the storage tank as a matter of safety (Fig. 22).

TEST YOURSELF

QUESTIONS – CHAPTER 1

1. Match the types of engines below with the average compression ratios.

   a. Gasoline engine   1. 16 to 1
   b. Kerosene engine   2. 8 to 1
   c. Diesel engine   3. 4 to 1

2. What happens in the cylinder of a gasoline engine during combustion "knock"?

3. Match the three pairs of items below:

   a. Gasoline Fuel 1. Octane number A. Must burn FAST
   b. Diesel fuel   2. Cetane number B. Must burn EVENLY

4. Which are more *volatile* – summer gasoline blends or winter blends?

5. In diesel engines, the fuel igniting too fast causes knocking.   True or False

6. Which is more volatile – diesel fuel with a cetane number of 40 or 60?

7. For gasoline fuel storage tanks that must be exposed to sunlight, which is the best color – grey, white or red?

8. An aboveground gasoline storage tank should be placed at least how many feet from the closest building?

   a.   12 ft. (3.6 m)
   b.   40 ft. (12 m)
   c   65 ft. (20 m)

9. LP-Gas storage tanks should be filled to what percent of their capacity?

   a.   80%
   b.   90%
   c.   95%

(Answers in the back of this book)

LUBRICANTS

Fig. 25 – Lubrication Protects the Precision Parts of the Modern Tractor

FUELS, LUBRICANTS, COOLANTS AND FILTERS

INTRODUCTION

Before modern machines, lubricants had but one purpose: to reduce friction and keep down wear between moving parts.

Selecting a lubricant was no particular problem; it was largely a matter of getting one heavy enough to maintain an effective film between the two contacting surfaces – one that would last from several hours to several weeks, depending on use.

As machines came into general use, the whole lubrication picture changed. The change has been particularly fast since 1940. The reason: loads are becoming heavier, speeds are faster, and moving parts fit much more closely than in earlier machines.

As a result, lubricants are constantly being changed and improved. As improvements are made, new terms are used such as "additives", "multi-viscosity", etc.

In the discussion that follows, we will explain those terms so you can better understand lubricants.

We will cover this information under the following headings:

1. **Engine Oils**

2. **Gear Oils**

3. **Differential Oils**

4. **Hydraulic and Transmission Fluids**

5. **Lubricating Greases**

ENGINE OILS

Once engine lubrication could be done by anyone with an oil can and single oil. Most oil was supplied in three grades – light, medium, and heavy.

But in today's high-speed engines, all this has changed. Everything is higher – compression, speeds, and temperatures.

Special oils have been developed for each type of engine, for each type of machine, and for each season.

As a result, lubrication has changed from a simple chore to a science of preventive maintenance.

WHAT ENGINE OILS MUST DO

Engine oils have several functions (Fig. 26). Here are four of the most important:

1. Oil reduces friction and wear.

2. Oil cools moving parts.

3. Oil helps seal the cylinders.

4. Oil keeps the parts clean.

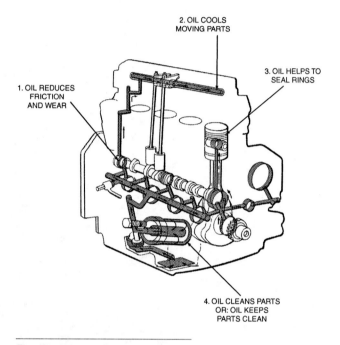

Fig. 26 – What an Engine Oil Must Do

Oil Reduces Friction and Wear

Scuffing and scoring is caused by metal-to-metal contact of moving parts. Wear also results from acid corrosion, rusting, and from the abrasion of contaminants carried in the oil.

To prevent metal-to-metal contact, the oil must maintain enough viscosity or thickness to provide a film or cushion between the moving parts under all operating temperatures. In spite of high internal heat, the viscosity must be no higher than necessary to give good starting and yet provide the least friction under sustained running.

Oil Cools Moving Parts

Engine oil is largely responsible for piston cooling. This is done by direct heat transfer through the oil film to the cylinder walls and on to the cooling system and by carrying heat from the underside of the piston crown and skirt to the engine crankcase.

Oils of equal viscosity have the same heat conductivity, but the oil must have enough heat stability to resist decomposition when in contact with these surfaces.

Oil Helps Seal the Cylinder

During combustion, pressures in the cylinder may be 1000 psi (7000 kPa) or higher. Oil helps the piston rings to seal these pressures in the cylinder by forming an oil film on the piston and cylinder walls.

Oil Keeps the Parts Clean

Contrary to popular opinion, engine oils do "wear out". Extended service not only depletes the additives, but also oxidizes the base oils to harmful compounds. While good filtration will prolong oil life, many contaminants are soluble in the oil and will pass through the filter. These contaminants are primarily unburned or partially unburned fuel, but corrosive acids and water are frequently present.

The oil must prevent the formation of these materials or, once formed, keep them in suspension so they do not settle inside the engine.

If an engine oil is to fulfill all these requirements, the oil must do the following:

1. *Keep a protective oil film on moving parts.*

2. *Resist high temperatures.*

3. *Resist corrosion and rusting.*

4. *Prevent ring sticking.*

5. *Prevent sludge formation.*

6. *Flow easily at low temperatures.*

7. *Resist foaming.*

8. *Resist breakdown after prolonged use.*

If we are to understand fully how engine oils perform their job and the differences between them, it is helpful to know how oils are produced.

PRODUCING TODAY'S ENGINE OILS

Today, in the United States alone, there are more than 300 oil refineries actively engaged in converting crude oil into gasoline, kerosene, fuel oils, lubricants and hundreds of other by-products. These refineries vary in size from those that process 150 barrels of crude oil a day to those that refine up to 160,000 barrels daily.

To understand better the differences in lubricating oils, lets take a look at one of these refineries and learn how petroleum or crude oil is converted to the specialized engine oils of today.

Fig. 27 – Distillation Plant for Distilling Crude Oil

Crude Oil is composed of thousands of different combinations of hydrogen and carbon called hydrocarbons. Crude oil, as it comes from the ground, also contains oxygen and traces of nitrogen and sulfur; sometimes clay, water, resins and mineral salts are present.

The various combinations of hydrogen and carbon give special characteristics to the "fractions" or parts of petroleum. Some of these fractions are valuable in themselves, such as natural gas, gasoline, and kerosene. Others must be changed through refining before they can be used. Separating these fractions and converting them to useful products are the two chief jobs of an oil refinery.

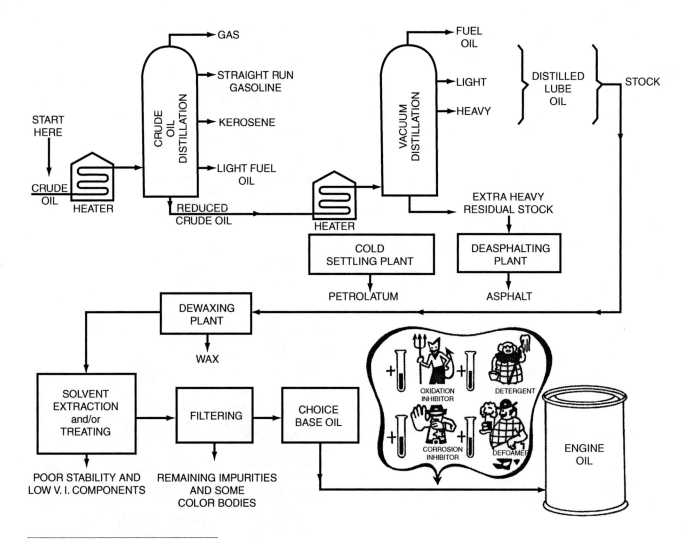

Fig. 28 – Lubricating Oil Refining Process

The key step in the refining process is **distillation** or "fractionation". It is by this process that the various parts of crude oil are separated from each other (Fig. 28).

Gases are removed from the top of the distillation tower. Light distillates such as gasoline are removed next, followed by middle distillates such as kerosene, light fuel oil, and diesel fuel oil.

Heavy fuel oils and lubricating oil base stocks are removed from the lower section of the tower. These heavier oils, often referred to as Reduced Crude Oil, are then further refined in the lube oil distillation columns, where steam and vacuum are provided to make distillation possible at a lower temperature.

In the lubricating oil still, the distilled lubricating oil base stock containing wax (but void of all asphalt) is removed from the top of the tower. This stock is further improved and refined by removal of wax, sulfur, sludge and other unwanted materials.

The lubricating oil base stocks are then blended to various viscosity. Various additives are introduced into the oil to make it flow more freely at low temperatures, to protect engine parts from rust and corrosion, and to give the oil special properties to adapt it to its special use.

Fig. 29 – Confused About Oil Ratings?

ENGINE OIL RATINGS AND CLASSIFICATIONS

To receive the best performance possible from a machine, lubricants of highest quality are required. The savings to be gained with lower quality lubricating oil are at best small in relation to the overall operating costs.

The apparent savings with cheap, inferior lubricating oils will be more than wiped out over the long run due to reduced output and greater maintenance costs. Therefore, be sure to use brands and grades of recognized quality...and deal only with suppliers having a reputation for dependability and the necessary technical skill to serve their needs.

How can you be assured of obtaining an oil of high quality and one that's right for the job? First, choose oil from a reliable supplier – reputable brand of oil. Then use the type of oil that's recommended for the equipment.

Several classifications of oil have been set up. These include:

1. *The SAE Viscosity established by the Society of Automotive Engineers.*

2. *API Service Classification established by the American Petroleum Institute.*

3. *MIL Specification prepared by the Ordnance Department of the U.S. Army, Navy and Air Force.*

4. *Manufacturer's Specifications developed by some engine makers.*

5. *ASTM Engine Sequence Tests whose procedures are adopted by the American Society for Testing Materials.*

Let's take a look at each of these rating systems.

SAE VISCOSITY

Oil **viscosity** is a measure of the fluidity of an oil at a given temperature. The lighter or more fluid oils are intended for winter use. Therefore, all W-grade oils are specifically tested under cold conditions to assure cold temperature performance.

In the viscosity test, a measured quantity of the oil is brought to the specified temperature. The length of time in seconds required for a specified volume of oil to flow through a small orifice in an instrument such as a Saybolt or Kinematic Viscometer is recorded. The SAE grade is determined by referring to the SAE (Society of Automotive Engineers) Chart below.

Oils vary in viscosity as temperatures change – becoming more fluid as temperatures increase and less fluid as temperatures decrease. The temperature effect on viscosity is not the same for all oils, and a measure of this is often important to the user.

SAE ENGINE VISCOSITY GRADES PER SAE J300				
SAE Viscosity Grade	Low Temperature Viscosity		Viscosity at 212°F ** (100°C)	
	Cold Cranking Temperature °F (°C)	Pumpability Temperature °F (°C)		
0W	3250 mPa·s at −22 (−30)	60 000 mPa·s at −40 (−40)	3.8 mm²/s	—
5W	3500 mPa·s at −13 (−25)	60 000 mPa·s at −31 (−35)	3.8 mm²/s	—
10W	3500 mPa·s at −4 (−20)	60 000 mPa·s at −22 (−30)	4.1 mm²/s	—
15W	3500 mPa·s at +5 (−15)	60 000 mPa·s at −13 (−25)	5.6 mm²/s	—
20W	4500 mPa·s at +14 (−10)	60 000 mPa·s at −4 (−20)	5.6 mm²/s	—
25W	6000 mPa·s at +23 (−5)	60 000 mPa·s at +5 (−15)	9.3 mm²/s	—
20	—	—	5.6 mm²/s	< 9.3 mm²/s
30	—	—	9.3 mm²/s	< 12.5 mm²/s
40	—	—	12.5 mm²/s	< 16.3 mm²/s
50	—	—	16.3 mm²/s	< 21.9 mm²/s
60	—	—	21.9 mm²/s	< 26.1 mm²/s

* Pumpability is a measure of the oil's ability to flow to the oil pump inlet and provide adequate oil pressure during start-up.
** All viscosity grades are tested for minimum viscosity at 212°F (100°C).

Multi-Grade Oils

Some oils are compounded to behave as light oils at cold temperatures, and as heavier oils at high temperatures. These oils are called **multi-grade** or **multi-viscosity** oils and include, for example, 10W-30 (Fig. 30). Where recommended, one multi-viscosity oil can replace as many as four or five single-grade oils.

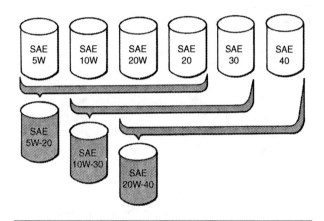

Fig. 30 – Multi-Viscosity Oil can Replace Several Single-Viscosity Oils – When Recommended

Multi-viscosity oils can give protection at both high and low temperatures. Initially, multi-viscosity oils were primarily intended for use during seasons in which both extreme cold and warm periods occur. They are now also recommended for use in warm weather in medium to heavy-duty applications.

Multi-grade oils are formulated by starting with a base oil of the lower viscosity grade such as 15W to which viscosity index improvers called polymers are added. These polymers do not significantly affect low temperature viscosity, but expand with increasing temperatures. This expansion causes an increase in viscosity at the higher temperature, yielding a multi-grade oil such as 15W-40.

Multi-grade oils such as 15W-40 may eliminate the need for seasonal changes of oil viscosity grade. The lower viscosity grade can lead to easier starting and improved fuel economy during warm up. The higher viscosity at high temperature controls oil consumption as well or better than the corresponding single grade.

Some tests have shown 15W-40 multi-grade oils reduced cylinder wear in engines that were run at partial load, but cylinder wear increased when the engines were run continuously at full load as compared to single viscosity 30 oil. This could be explained by recognizing that the cylinder temperature at full load was high enough to cause partial evaporation of the multi-grade oil while at partial load it was not.

SAE viscosity numbers are widely used as a means of identifying and classifying lubricating oil, and today are found on practically all cans or drums of oil marketed in the world (Fig. 31).

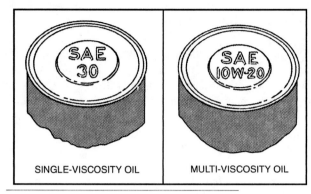

Fig. 31 – Oil Viscosity is Marked on Oil Containers

Each number indicates permissible range or limit of viscosity at specified temperatures.

No attempt is made to define the oil's quality, additive content, performance value or suitability for specified service conditions.

Therefore, the SAE viscosity should not be the only factor considered when selecting an oil, but the service rating of engine oils classified by the American Petroleum Institute (API) is also important.

Arctic Engine Oils

These oils are relatively new and are designated 0W-20 or 0W-30. Operating in extreme cold temperatures requires special cold-weather operating procedures. Always consult the engine manufacturer's recommendations for operating in arctic-like conditions and recommended oil viscosity grades for certain temperature ranges.

Temperature versus Viscosity

Manufacturers normally recommend engine oil viscosity grades for various operating temperatures (Fig. 32). Use only the oil viscosity recommendations found in the operator's manual for a specific machine. Do not assume that similar machines have the same oil viscosity recommendations. Certain engines and other machine design characteristics may require different viscosity grades to perform well.

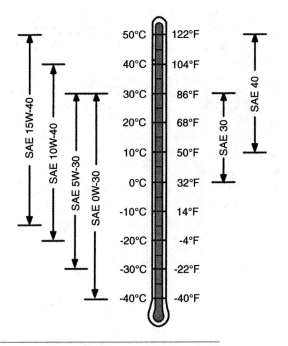

Fig. 32 – Typical Diesel Engine Oil Viscosity Chart

API SERVICE DESIGNATIONS

The API classification system is a joint effort of the American Petroleum Institute (API), American Society for Testing and Materials (ASTM), and Society of Automotive Engineers (SAE) organizations. It attempts to clarify the oil specification and better define oil qualities between the engine manufacturer, the petroleum industry, and the customer.

Often on farms and on certain types of construction jobs where diesel-powered equipment is operated along with gasoline-powered equipment, operators may wish to lubricate all of their equipment with the same type of oil. In such cases, a diesel engine oil that can also be used in gasoline engines may be desired. Lubricants meeting more than one classification can be so marked. Some oils, therefore, are designated "For Service CE/SG" or "For Service CC/SG".

API oil classifications are shown in the "API Service Ratings for Engine Oils" chart shown on the next few pages.

The most important part of selecting an engine oil is using the equipment manufacturer's recommendations found in the operator manual. For example, the recommended engine oil for a current gasoline engine used in an automobile is for service category SG (see chart). In some cases the manufacturer may recommend an oil that has two classifications – CC/SG. If the operator uses an oil with only one of these classifications, the warranty may be voided and, worse, the engine may not get the protection it requires.

Expensive mistakes may also occur when substituting an SD category oil for an SG category or a CC oil for a CE, etc.

PERFORMANCE TESTS

The American Society for Testing and Materials (ASTM) has identified several performance tests to evaluate diesel and gasoline engine oils. Some of the common tests are described in the following chart. The test numbers identify the condition measured. Tests numbered II, III and V have been updated several times. Prior tests for the same condition carried A, B or C suffixes.

Fig. 33 – Research Scientists Examine Engine Parts after API Service Tests

DIESEL VERSUS GASOLINE ENGINE OILS

Oil requirements for diesel engines differ substantially from those for gasoline engines primarily due to the operating temperatures and conditions of use. The sulfur content of diesel fuel is also a problem. Diesel engine oils must help protect against the formation of sulfuric acid that causes corrosion.

Diesel engines are normally used in heavier equipment and subjected to higher load factors for more extended periods than gasoline engines used in automobiles. In general terms, requirements are very similar; however, additives that perform acceptably in one type of engine may not perform to the same degree in the other type of engine.

The military engine oil specifications should be used only as a yardstick for helping to assure products are suited for the purpose intended. They should not be used to tell the optimum or ultimate in performance. These specifications describe products which will work with a reasonable degree of satisfaction for a given requirement, but they do not necessarily indicate that the product is the best quality oil or the least expensive available.

Although military specifications do not govern non-military uses for engine oils, many non-military users request oils which have qualified under these requirements because such oils have had successful performance records. For this reason, oil manufacturers often identify their containers with the MIL specification.

Today many machine makers have gone beyond the standards and have developed their own specific oils. An example is "Torq-Gard Supreme® Plus-50" developed specifically for John Deere engines. This oil has been formulated for off-the-road operation as opposed to road driving and gives you added advantages not offered by automotive engine oils.

Fig. 34 – Engine Oils are Highly Refined Petroleum

API SERVICE RATINGS FOR ENGINE OILS

Letter Designation	API Brief Identification and Engine Service Description	ASTM Engine Oil Description
SB**	*Minimum Duty Gasoline Engine Service* Service typical of engines operated under such mild conditions that only minimum protection afforded by compounding is desired. Oils designed for this service have been used since the 1930's and provide only anti-scuff capability, and resistance to oil oxidation and bearing corrosion.	Provides some anti-oxidant and anti-scuff capabilities.
SC**	*1964 Gasoline Engine Warranty Service* Service typical of gasoline engines in 1964 through 1967 models of passenger cars and trucks operating under engine manufacturers' warranties in effect during those model years. Oils designed for this service provide control of high and low temperature deposits, wear, rust and corrosion in gasoline engines.	Oil meeting the 1964-1967 requirements of the automobile manufacturers. Intended primarily for use in passenger cars. Provides low temperature anti-sludge and anti-rust performance. *Tests required:* IIA, IIIA, V, L-38*
SD**	*1968 Gasoline Engine Warranty Service* Service typical of gasoline engines in passenger cars and trucks (1968 through 1970 models) and operating under engine manufacturers' warranties. Oils designed for this service provide more protection from high and low temperature engine deposits, wear, rust and corrosion in gasoline engines than oils for API Service Classification SC and may be used when oils for API Service Classification SC are recommended.	Oil meeting the 1968-1971 requirements for automobile manufacturers. Intended primarily for use in passenger cars. Provides low temperature anti-sludge and anti-rust performance. *Tests required:* IIB, IIIB, V-B, L-38*
SE**	*1972 Gasoline Engine Warranty Service* Service typical of gasoline engines in passenger cars and trucks beginning with 1972 and certain 1971 models and operating under engine manufacturer's warranties. Oils designed for this service provide more protection against oil oxidation, high-temperature engine deposits, rust and corrosion than SC or SD and may be used when oils for API Gasoline Engine Warranty Maintenance Classification SC or SD are recommended.	Oil meeting the 1972 requirements of the automobile manufacturers. Intended primarily for use in passenger cars. Provides high temperature anti-oxidation, low temperature anti-sludge and anti-rust performance. *Tests required:* IIC, IIIC, V-C, L-38*
SF**	*1980 Gasoline Engine Warranty Service* Service typical of gasoline engines in passenger cars and some trucks beginning with the 1980 model and operating under engine manufacturer's recommended maintenance procedures. Oils developed for this service provide increased oxidation stability and improved anti-wear performance relative to oils that meet the minimum requirements for API Service Category SE. These oils also provide protection against engine deposits, rust and corrosion. Oils meeting API Service Category SF may be used where API Service Categories SE, SD or SC are recommended.	Oil meeting the 1980 warranty requirements of the automobile manufacturers. Intended primarily for use in gasoline engine passenger cars. Provides protection against sludge, varnish, rust, wear, and high-temperature thickening. *Tests required:* IID, IIID, V-D, L-38*

API SERVICE RATINGS FOR ENGINE OILS - Continued

Letter Designation	API Brief Identification and Engine Service Description	ASTM Engine Oil Description
SG	*1989 Gasoline Engine Warranty Service* The category SG denotes service typical of present gasoline engines under manufacturers' recommended maintenance procedures. Category SG quality oil includes the performance properties of API Service Category CC. (Certain manufacturers of gasoline engines require oils also meeting the higher diesel engine Category CD.) Oils developed for this service provide improved control of engine deposits, oil oxidation, and engine wear relative to oil developed for previous categories. Oils meeting API Service Category SG may be used when API Service Categories SF, SE, SF/CC or SE/CC are recommended.	Oil meeting the 1989 warranty requirements of the automobile manufacturers. Intended primarily for gasoline engine passenger cars but will also qualify for moderate diesel engine service. Provides protection against rust, oil thickening, valve train wear, sludge, piston varnish, and piston deposits. *Tests required:* IID, IIIE, V-E, L-38; 1H2
SH	*1994 Gasoline Engine Warranty Service* This oil is for use in service typical of gasoline engines in current and earlier passenger car, van, and light truck operation under vehicle manufacturers' recommended maintenance procedures. Category SH oils provide improved control of engine deposits, oil oxidation, engine wear, rust and corrosion.	Oil meeting the 1994 warranty requirements of the automobile manufacturers. Provides protection against rust, oil thickening, valve train wear, sludge, piston varnish, and piston deposits. *Tests required:* IID, IIIE, V-E, L-38*
SJ	*1997 Gasoline Engine Warranty Service* This oil is for use in service typical of gasoline engines in passenger car, van, and light truck operation under vehicle manufacturers' recommended maintenance procedures. Category SJ oils provide improved control of engine deposits, oil oxidation, engine wear, rust and corrosion.	Oil meeting the 1997 warranty requirements of the automobile manufacturers.
CA**	*Light Duty Diesel Engine Service* Service typical of diesel engines operated in mild to moderate duty with high quality fuels. Occasionally has included gasoline engines in mild service. Oils designed for this service were widely used in the late 1940's and 1950's. These oils provide protection from bearing corrosion and from high temperature deposits in normally aspirated diesel engines when using fuels of such quality that they impose no unusual requirements for wear and deposit protection.	Oil meeting the requirements of MIL-L-2104A. For use in gasoline and naturally aspirated diesel engines operated on low sulfur fuel. The MIL-L-2104A specification was issued in 1954.
CB**	*Moderate Duty Diesel Engine Service* Service typical of diesel engines operated in mild to moderate duty, but with lower quality fuels that necessitate more protection from wear and deposits. Occasionally has included gasoline engines in mild service. Oils designed for this service were introduced in 1949. Such oils provide necessary protection from bearing corrosion and from high temperature deposits in normally aspirated diesel engines with higher sulfur fuels.	Oil for use in gasoline and naturally aspirated diesel engines. Includes MIL-L-2104A oils where diesel engine test was run using high sulfur fuel.

API SERVICE RATINGS FOR ENGINE OILS - Continued

Letter Designation	API Brief Identification and Engine Service Description	ASTM Engine Oil Description
CC**	*Moderate Duty Diesel & Gasoline Engine Service* Service typical of lightly supercharged diesel engines operated in moderate to severe duty and has included certain heavy-duty, gasoline engines. Oils designed for this service were introduced in 1961 and used in many trucks and in industrial and construction equipment and farm tractors. These oils provide protection from high temperature deposits in lightly supercharged diesels and also from rust, corrosion and low temperature deposits in gasoline engines.	Oil meeting the requirements of MIL-L-2104B. Provides low temperature anti-sludge, anti-rust and lightly supercharged diesel engine performance. The MIL-L-2104B specification was issued in 1964. *Tests required: IIB, IH2*
CD**	*Severe Duty Diesel Engine Service 1* Service typical of supercharged diesel engines in high-speed, high-output duty requiring highly effective control of wear and deposits. Oils designed for this service were introduced in 1955, and provide protection from bearing corrosion and from high temperature deposits in supercharged diesel engines when using fuels of a wide quality range.	Oil meeting the requirements of the Caterpillar Tractor Co. Series 3 specification. Provides moderately supercharged diesel engine performance. Caterpillar established the Series 3 specification in 1955. The related MIL-L-45199 specification was issued in 1958. *Tests required: IG2, L-38*
CE**	*Severe Duty Diesel Engine Service 2* Service typical of turbocharged or supercharged heavy duty diesel engines manufactured since 1983 and operated under both low-speed, high-load, and high-speed, high-load conditions. Oil designated for this service may also be used when previous API engine service categories for diesel engines are recommended.	Oil meeting the performance requirements described in Category CD by the IG2 and L-38 tests plus the Mack Trucks, Inc. T-6 and T-7 tests (1984) and the Cummins Engine Co. NTC-400 tests (1983) to address oil consumption, deposits, wear, and oil thickening. Provides turbocharged, direct injection diesel performance.
CF-4	*Severe Duty Diesel Engine Service 3* Service typical of high-speed turbocharged or supercharged four-stroke cycle diesel engines. CF-4 oils are particularly suited for on-highway heavy duty truck engines. CF-4 oils exceed the requirements of the CE category oils. CF-4 oils will ultimately replace CE category oils. CF-4 oils may also be used in place of earlier CC and CD oils.	Oil meeting performance requirements under test techniques T-6, T-7 and NTC-400, but with greater demand on improved oil consumption control and reduced piston deposits.
CG-4	*Severe Duty Diesel Engine Service* Service typically used in highway and off-highway applications where the fuel sulfur may vary from less than 0.05 percent by weight to less than 0.5 percent by weight.	These oils are especially effective in engines designed to meet 1994 exhaust emission standards and may also be used in engines requiring API Service Categories CD, CE and CF-4. Oils designated for this service have been in existence since 1995.
CH-4	*Severe Duty Diesel Engine Service* Service typically used in highway and off-highway applications where the fuel sulfur may vary from less than 0.05 percent by weight to less than 0.5 percent by weight.	These oils are designed to provide protection for diesel engines meeting 1998 on highway emission standards. It may be used in engines requiring API Service Categories CD, CE, CF-4, and CG-4. Oils designated for this service have been in existence since 1998.

*Note: For a description of these tests, see the chart "ASTM Performance Tests" previously shown. **Obsolete Non-Current Categories.

CONTAMINATION OF OIL

Most people – even many service technicians—are still not fully aware of the harmful effect of **contaminants** on engine oil's performance. Contaminants seriously hamper good lubrication, regardless of the oil's original quality.

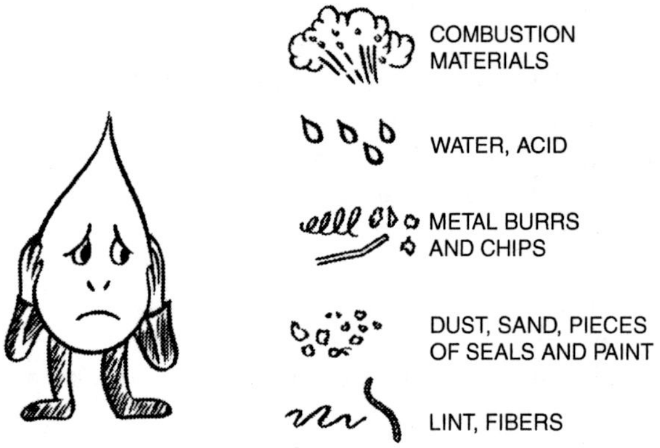

COMBUSTION MATERIALS

WATER, ACID

METAL BURRS AND CHIPS

DUST, SAND, PIECES OF SEALS AND PAINT

LINT, FIBERS

Fig. 35 – Contamination of Oil

Engine oil quality is probably being stressed more today than ever before, but the effect of oil contaminants many times hinders engine performance even more than using a poor quality oil. As a matter of fact, contaminants that collect in oil under normal operating conditions, to say nothing of adverse conditions, can reduce engine life more than any other single factor.

Let's take a look at some of the many oil contaminants, how they enter the engine, what oil companies are doing to counteract them and what you can do.

Foreign Particles in Oil

Probably the most familiar of the external contaminants is dust breathed in with the combustion air (Fig. 36).

Fig. 36 – Dust is a Major Source of Oil Contamination

Similar material also enters the engine crankcase directly as a result of the breathing action taking place there.

In diesels, fuel soot particles from combustion enter the crankcase oil with blowby gases.

Microscopic metal particles also get into the oil as a result of normal engine wear.

As these foreign particles accumulate, increased wear soon results in the cylinder bore, on piston rings, and within bearings, even though the best oil was used originally. Early engine failure may result.

These tiny particles also restrict oil flow, and in combination with water and oxidized products, form sludge. Metal particles accelerate wear and damage bearings. Fuel soot thickens the oil and interferes with lubrication.

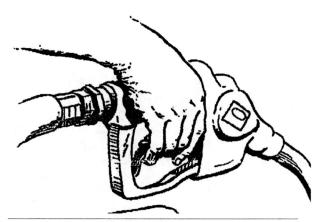

Fig. 37 – Pre-Cleaners and Pre-Screeners Keep Large Foreign Particles from Reaching the Air Cleaner

Oil companies have helped in the fight against these contaminants by introducing additives into the oil. **Anti-scuff** additives help reduce the number of metal particles resulting from engine wear. **Detergents** reduce deposit build-up. **Dispersants** keep contaminants finely dispersed in the oil, thus not interfering with the lubricating qualities of the oil. This also permits particles to drain out with the oil.

Machine operators can also help in preventing early engine failure through proper care of the air filter elements by soaking in a filter element cleaner/water solution (Fig. 38-A). Or by using an element cleaning gun (Fig. 38-B), oil filler breather cap, and crankcase ventilator...through regular and frequent oil filter changes...and by proper storage and handling of lubricants.

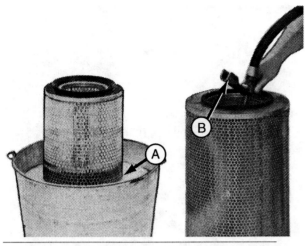

Fig. 38 – Regular Filter Cleaning can Prevent Costly Engine Damage

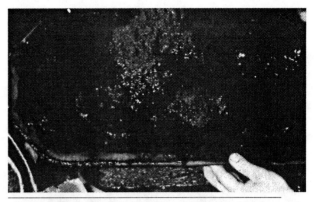

Fig. 39 – Ice in Crankcase Following Cold Engine Operation

Lubricants should be stored in clean, closed cabinets or rooms; covers and pour spouts on the drums or containers should be kept closed when not in use. This not only keeps out impurities, but it also reduces condensation of water caused by atmospheric changes. This will be covered later under "Storage and Handling".

Water Contamination of Oil

Water vapor, a normal product of combustion, tends to condense and collect as water in the crankcase during periods of cold engine operation. Each gallon of fuel consumed produces more than a gallon of water. This water, which is not completely vaporized until the cylinder wall temperature reaches 145°F (63°C), condenses on colder cylinder walls and is scraped down into the crankcase oil by the piston ring.

Water is a troublemaker that causes rusting of vital steel and iron surfaces. Eight times as much engine or cylinder wear occurs in an engine operating at 100°F (38°C) than at 160°F (71°C).

Water in the crankcase may freeze (Fig. 39) or it may combine with oxidized oil and carbon to form cold-engine sludge. Sludge can plug oil screens and often collects in piston oil ring grooves, interfering with ring action. Oil line plugging also occurs at times. Rust, which may result from water, further contaminates the oil.

Oil companies have introduced special rust inhibitors into oils. These inhibitors aid in protecting precision parts. Detergent-dispersants keep parts clean and deposits in suspension, helping to prevent sludge.

You can help prevent severe engine damage by:

1. *Warming up engine properly before a load is applied.*

2. *Making sure engines are brought up to normal operating temperatures each time they are used.*

3. *Using proper thermostats to heat engines to correct temperatures as quickly as possible.*

4. *Checking engine temperatures frequently.*

5. *Draining crankcase oil while engine is warm.*

Antifreeze Contamination of Oil

Proper cooling system maintenance will prevent contamination of engine oil by antifreeze.

Antifreeze will oxidize in oil, forming sludge or gummy, resinous products (Fig. 40). It also tends to strip out other oil additives so they no longer give protection. This results in a varnish build-up on bearings. Sludge forms on the oil screen, oil pan and in push rod chambers while rust forms on the valve train. Antifreeze contamination usually results in the need for a complete engine overhaul. Special precautions can be taken against antifreeze contamination by:

1. *Following specified service manual procedures when torquing head bolts during overhaul. (NOTE: Be sure to retorque bolts, when specified).*

2. *Using a cooling system sealer prior to filling with permanent antifreeze, if approved.*

3. *Repairing all leaks immediately. Not depending on a stop-leak as a permanent fix.*

4. *Guarding against incorrect timing and improper use of starting fluids in diesel engines, both which can result in head gasket damage.*

Fig. 40 – Antifreeze in Crankcase Results in Sludge Formation and can Clog Oil Screens

Fuel Contamination of Oil

In **gasoline engines**, over-choking, engine missing, carburetor flooding, and cold engine operation all tend to let gasoline seep into the oil. This raw fuel runs down the cylinder walls, past the rings, washing away the lubricating oil and increasing engine wear.

In **diesel engines**, fuel contamination of oil may be caused by a cracked fuel pump diaphragm or by a leak in the fuel injection pump shaft seal. Piston seizure and short bearing life can result in diesel engines if crankcase dilution is excessive.

In both types of engines (diesel and gasoline), the partially oxidized fuel and the unburned fuel which mix with oil in the crankcase contribute to deposits forming in engines in general and on piston surfaces (as varnish) in particular.

Fig. 41 – Varnish Build-up on Pistons

IMPORTANT: Deposits on pistons cause rings to stick (Fig. 41), and accelerated engine wear usually results.

Antioxidants are added to some oils to help halt the chemical processes that result in varnish or acid formation. Detergent-dispersants are added to carry any varnish or deposits out of the ring area and to keep them in suspension until oil is drained.

Machine owners can help prevent fuel contamination of oil by remembering these simple rules:

1. *Don't over-choke an engine.*

2. *Never run an engine when it is missing.*

3. *Avoid too much idling of diesel engines.*

4. *Have flooding carburetors repaired.*

5. *Keep the diesel fuel system in good operating condition.*

6. *Make sure engines are brought up to normal operating temperatures each time they are used.*

7. *Buy clean-burning fuel.*

Overheating of Oil

High, operating temperatures, caused by heavy loads, faulty cooling systems, bad timing, and pre-ignition and detonation speed up oxidation of oils. A rule of thumb: every 18°F increase in temperature above 185°F (85°C) **doubles** the oxidation rate of average oil.

Oxidation breaks down the oil and forms deposits. This results in ring sticking, valve sticking, and sludge. The antioxidants help protect the oil from oxidation and reduce oil breakdown.

You can help prevent engine damage from oxidation due to high operating temperatures by making sure the cooling system is properly maintained and that temperature gauges are working properly. Check the engine temperature frequently...have the engine timing checked periodically (Fig. 42)...and use proper fuel as recommended in the operator manual. To get better engine performance, lower fuel consumption and less overheating, avoid over-working the engine.

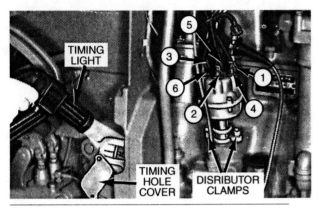

Fig. 42 – Have the Engine Tuned as Outlined in the Operator Manual

Oxidation of Oil

In the crankcase, oil is sprayed from the various moving parts, forming a hot mist of oil particles and air.

In the combustion chamber, the oil is also partly a mist, because it is scraped up the cylinder walls at very high speeds by the rising piston rings, and heated.

Oxidation products are formed in the oil. Their nature depends on the temperature.

In the crankcase, oxidation produces acids and the complex carbonaceous products known as "asphaltenes". These materials, in combination with fuel contaminants, assist in forming the stable sludge that is particularly related to low-temperature operation.

In the combustion chamber, oil forms a so-called carbon deposit by combination of oxidation and decomposition by heat. Part of this deposit is washed back to the sump while part remains in the combustion space.

The effects of oxidation will thus aggravate the condition of the oil caused by contamination with the combustion products. The result will be an oil containing acids which may corrode bearing metals and also form resins which may deposit on the pistons and hot metal parts as varnish.

The use of highly refined oils has lessened the risk of oxidation, and anti-oxidant additives can obtain further improvement. It is also possible to include either additives to neutralize the acids formed by combustion of the fuel or, alternatively, one additive acting as anti-oxidant and neutralizer. Such additives can greatly improve the performance of engine oils and are used today in most grades.

Oil Additives

Special oil additives are put into lubricating oils to provide the extra performance required of today's high-speed engines.

Each additive, or combination of additives, is included in oil for a specific reason, based on the service expected for that oil.

Oil may contain none, some, or all of the additives described below:

Note: Most quality engine oils already contain the needed additives for all conditions where they are recommended. Therefore, most engine manufacturers say "NEVER PUT ADDITIVES IN THE ENGINE CRANKCASE." The use of additional crankcase additives may actually reduce oil protection rather than help it.

1. Anti-Corrosion Additive
   Helps prevent failure of alloy bearings from corrosive acids that are formed as a normal by-product of combustion. Protects other metal surfaces from corrosive attack. Works with oxidation inhibitors.

2. Oxidation Inhibitor Additive
   Helps keep oil from oxidizing even at high temperatures. Prevents acid, varnish and sludge formations. Protects alloy bearings from corrosion. Prevents oil molecules from combining with oxygen.

3. Anti-Rust Additive
   Prevents rusting of metal parts during storage periods, downtime, or even overnight or weekends. Neutralizes acids so they are no longer harmful. Clings to metal surfaces and builds up a protective coating, which repels water droplets and protects metal from rust.

4. Viscosity Index Improver

Helps an oil give top-lubricating protection at both low and high temperatures. Multi-grade oils with this additive span an extra-wide range of viscosity grades compared to single grade oils. Lighter oils make starting easier at lower temperatures, but thin out as the oil heats up. Heavier oils give good protection at high temperatures, but become thick and cause hard starting and improper lubrication at low temperatures. A viscosity index improver gives the oil the beneficial properties of both light and heavy oils.

5. Pour Point Depressant Additive

Prevents wax crystals from congealing in cold weather and forming clumps. (Some paraffin wax is present in lubricating oil, although most wax is removed during refining processes). These wax crystals tend to congeal in cold weather, interfering with oil flow and causing lubrication difficulties.

6. Extreme Pressure Additive

Assures lubrication where extreme pressures between close tolerance and metal-to-metal surfaces are encountered. They reduce friction, prevent galling, scoring, seizure and wear.

7. Detergent-Dispersant Additive

Helps keep metal surfaces clean and prevents deposit formation. Particles of soot and oxidized oil or fuel are kept suspended in the oil. Suspension is so fine that it passes through oil filter and continues to be carried by the oil. Regular drain periods result in the removal of suspended contaminants. Black oil is evidence that oil is helping keep the engine clean by carrying combustion particles in the oil rather than letting them accumulate on parts as sludge.

8. Foam Inhibitor Additive

Helps prevent air bubbles that would otherwise restrict lubrication. Speeds up rate at which air bubbles break up, keeping oil from foaming as it circulates rapidly in today's modern, high-speed engines.

Facts about Oils

Many people hold mistaken ideas about oils. Some of these people learn the hard way – through expensive overhauls. Let's look at some facts about oils.

1. Oil Does Wear Out

Some people believe oil never wears out. This is a mistaken belief of many that feel oil is always slippery and therefore is always effective. Oil loses many of its good lubricating qualities as it absorbs contaminants and as its additives are depleted. Acid formation, sludge and varnish, and engine deposits all tend to cause the oil to become unfit for further use (Fig. 43).

Fig. 43 – Always Use New Oil in the Engine Crankcase

2. Black Oil May Not Require Replacement

For example, in a diesel engine, oil *should* turn black with use to be effective. Detergent and dispersant additives in the oil clean and hold deposits in suspension for removal at the time oil is drained.

3. Cheap Oil Does Not Save Money

Too many people today are trying to save money on oil, when they should be saving their engines. Ten dollars saved through cheap oils can result in hundreds of dollars of expense through engine repairs.

4. Use Only the Correct Filter

Any filter is better than a clogged filter, but the money saved by buying cheap, "will-fit" filters is negligible compared to the engine damage that can possibly result from their use. Some filters are so dense that not enough oil will pass through. This causes most of the oil to bypass the filter. Others however, permit damaging contaminants to pass through. Buy quality oil filters as recommended in your machine operator manual (Fig. 44).

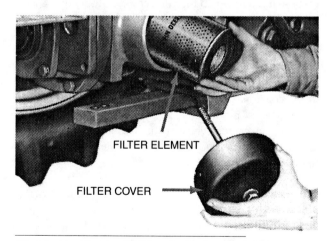

FILTER ELEMENT

FILTER COVER

Fig. 44 – Use Genuine Oil Filters for Replacement

5. Oil Oxidation Results in Thicker Oil
   Oil oxidation results in **thicker** oil. A viscosity increase indicates that either oil is breaking down or it is badly contaminated. On the other hand, two things cause thinning of oil – high temperatures and dilution by fuel.

6. Light Oils are Not Best
   Many operators like to use a light oil as long as they can, and switch to heavier oils as oil consumption increases. This is not the best practice to follow. Use the chart in your operator manual to insure use of the correct oil. Oil with the right viscosity is essential to good lubrication.

7. Operator Manual Recommendations are Critical
   Many owners pay far too little attention to their operator manuals. Lubrication recommendations in these manuals are there to insure good performance over a long period of time. Refer to the operator manual and follow the recommended lubrication practices (Fig. 45).

Fig. 45 – Read Your Operator Manual and Use it as a Service Reminder

8. Synthetic Oils are Not Necessarily Better
   Synthetic oils may last longer than petroleum based oils in passenger cars, as advertised. But, no experts say synthetic oils are categorically better than petroleum based oils. Also, there has been no statement by experts stating synthetic oil is better for industrial and agricultural engines.

Special Oils for Off-The-Road Machines

With the development of modern high-compression, high-speed engines with very close tolerances, it became apparent to some engine manufacturers that modern day oil did not fully satisfy the requirements of some engines.

Most modern oils were compounded basically to satisfy a wide range of requirements – automobile, truck, tractors, industrial, and aviation engines.

They had to withstand, particularly with automobiles and trucks, a great deal of stop-and-go operation and at the same time perform satisfactorily under continuous, full-load operation as in farm and industrial machine applications.

Initially, automobile and truck oil change recommendations were at 1000-mile (1600-kilometer) intervals. Later the recommendation was changed to 3000-mile (4800-kilometer) intervals or every 60 days, whichever occurred first. It is now up to 6000-mile (9600-kilometer) intervals.

Under the best operating conditions, which would be continuous day-by-day driving at an average speed of 50-mph (80 km/h), 3000 miles (4800 kilometers) of driving would take 60 hours. Stop-and-go, excessive idling and cold engine operation imposed greater demands for high-quality oil.

Farm and industrial machines, however, present a different picture. They normally operate continuously under full load, at a constant speed, over relatively long periods of time, with a minimum of stop-and-go operation. Oil change periods vary; therefore you must refer to your operator manual for the proper oil change intervals.

Compared to automobile and truck operation, it is obvious that farm and industrial machine engines present a much different lubrication problem. For example, automobile and truck oil changes are recommended after 6000 miles (9600 kilometers) of travel or about 120 hours at an average of 50-mph (80 km/h). Farm and industrial engine oil changes are typically recommended every 100-250 hours of operation. Converted to miles traveled at 50 mph (80 km/h), every 100 hours would represent **5000 miles (8000 kilometers)** of automobile travel.

Fig. 46 – Scored Piston – Due to Loss of Oil Film

While it is true that the automobile engine is subjected to much more stop-and-go, idling, and cold engine operation, it is seldom subjected to extended, full-load operation.

The farm and industrial engine has the advantage of less stop-and-go operation and of operating most of the time at normal operating temperatures. But they have the disadvantage of operating mostly under the high pressures and combustion heat of full-load operation.

Engineers who studied the problem noted that engine oils compounded basically for automobile and truck engines. However, their use left something to be desired when used in off-the-road machines.

Fig. 47 – Engine Bearing Corroded by Acid Formation

To overcome this problem, the engineers worked with oil companies to develop oils that would perform better in farm and industrial engines.

The result was the development of superior quality oil that performed very satisfactorily in off-the-road machines.

Engine Oil Service Tips

These recommendations will help assure the longest possible life for an engine:

1. *Check oil level daily (Fig. 48).*

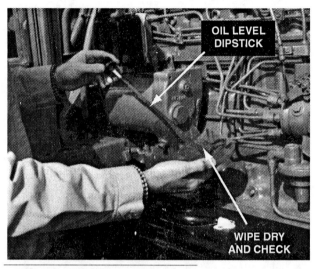

Fig. 48 – Check Oil Crankcase Level Daily

2. *When adding or changing oil, be sure the correct service classification is used. Clean oil containers and fill spout (Fig. 49).*

Fig. 49 – Clean Oil Container and Filling Equipment

3. *Replace the oil filter when recommended. Use only the specified filter; do not use a substitute (Fig. 50).*

Fig. 50 – Replace Oil Filter with Only the Specified Filter

4. Check for leaks after changing oil and oil filter (Fig. 51).

Fig. 51 – Check for Leaks After Changing Oil

NOTE: For more on engines, see the FOS "Engines" manual.

Fig. 52 – Check the Engine Oil Level Daily

Storing and Handling Engine Oils

Use care in storing and handling oil to keep out dirt and moisture.

Store oil inside whenever possible – in clean, closed cabinets or rooms. It is best to keep oil relatively warm in the winter. Figure 53 shows how changes in temperature can draw water into oil barrels.

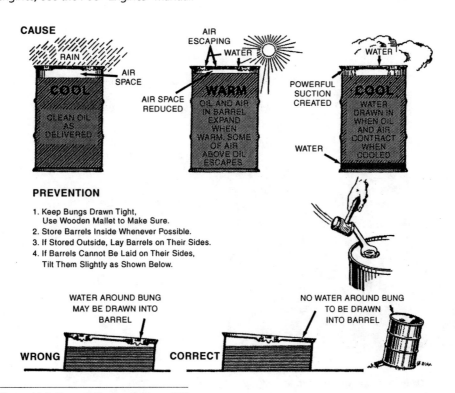

Fig. 53– Storage Practices which Prevent Contamination of Oil

If stored outside, lay oil barrels on their sides. If they cannot be laid on their sides, tilt them slightly as shown in Fig. 53; turn the barrel so that the bung is at the high side, away from any water which collects.

Keep oil bungs drawn tight. Use wooden mallets to make sure.

Rinse oil containers and funnels in fuel after use. Cover them to keep out dirt, or store them upside down.

When adding oil, clean all dirt from around the filler cap before removing it. (Do the same before unscrewing an oil filter or a filter cap).

Dispose of Oil Properly

Improper disposing of drained oil can harm the environment and ecology. Never pour oil on the ground, down a drain or into a waterway, stream, lake, or pond (Fig. 54).

Always observe environmental regulations.

Fig. 54 – Always Dispose of Oil Properly

GEAR OILS

Gear oils are those used in enclosed gearboxes to lubricate mechanical transmissions, differentials, and steering gears.

Fig. 55 – Transmission-Differential Unit which Operates in Oil Bath

Today's high-speed, high-torque power trains use relatively small gears. The result is high tooth loads and harder rubbing between mating gears. This makes lubrication more critical. Unless a good film of fluid is maintained between the mating gear teeth, they will wear and score (Fig. 56).

NORMAL WEAR

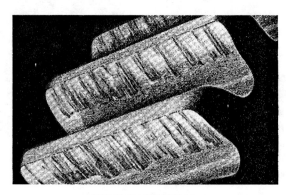

SCORING WEAR

Fig. 56 – Normal Gear Wear and Scoring Wear

GEAR LUBRICATION

In small, simple gear units (Fig. 57) where tooth pressures are relatively light, simple gear oil is often adequate.

Special gear arrangements such as spiral bevel and worm gears can wear very rapidly unless the correct gear oil is used.

For these higher loads, the gear oil must have special "anti-friction" and "anti-weld" agents. Oils containing these additives are known as **extreme pressure** or **hypoid** lubricants.

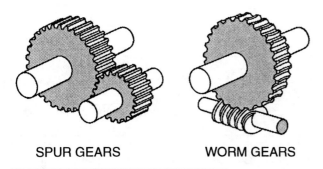

SPUR GEARS   WORM GEARS

Fig. 57 – Gear Oils Must Match the Gear Load

GEAR OIL REQUIREMENTS

To perform satisfactorily under today's conditions, most gear oils should have the following properties:

Extreme Pressure Properties

Extreme pressure properties are required in gear systems where hypoid, heavily loaded spiral-bevel and worm gear combinations are used. Spur gears and moderately loaded gears do not require this property.

Oxidation Stability

Gear lubricants must be chemically stable to resist oxidation and sludge formation under sustained heat with violent agitation and air foaming.

Corrosion Resistance

Extreme pressure agents are, by nature, chemically active and protect gear teeth by coating them. Discoloration of gears and internal parts frequently occurs, but this does not indicate abnormal corrosion.

Foam Resistance

Foam resistance is mandatory in gear lubricants because of the violent agitation of the oil. Most gear lubricants contain foam suppressors.

Viscosity Index

The viscosity index indicates the change in viscosity with changes in temperature. The higher the viscosity index, the less impact temperature has on viscosity. Because of the wide variation in both ambient and gear case temperatures, a higher viscosity index is desirable.

Pour Point

The pour point must be low enough to provide lubrication at the lowest anticipated temperature. Low viscosity gear oils normally have a low pour point – down to -50°F (-46°C).

Channel Temperature

Gear oils should be fluid enough at the lowest operating temperature to flow and cover moving parts, rather than to form a channel in which the gears can move free of lubricant. Channel temperatures may be as much as 15°F (-8°C) colder than the specified pour point of the lubricant.

SAE GEAR CLASSIFICATION

Society of Automotive Engineers (SAE) gear oil classification is based on viscosity alone and is no indication of quality or service (see Axle and Manual Transmission Lubricant Viscosity Classification Chart).

AXLE AND MANUAL TRANSMISSION LUBRICANT VISCOSITY CLASSIFICATION CHART			
SAE Viscosity Grade	Maximum Temperature For Viscosity of 150 000 cP °C	Viscosity at 100°C cSt	
		Minimum	Maximum
70W	-55	4.1	—
75W	-40	4.1	—
80W	-26	7.0	—
85W	-12	11.0	—
90	—	13.5	<24.0
140	—	24.0	<41.0
250	—	41.0	—

While the SAE numbers of gear oils are higher than the SAE numbers of engine crankcase oils, gear oils are not necessarily that much higher in viscosity. To avoid confusion, higher numbers are assigned to gear oils (Fig. 58). For example, SAE 80W gear oil actually has about the same viscosity as SAE 20W engine oil.

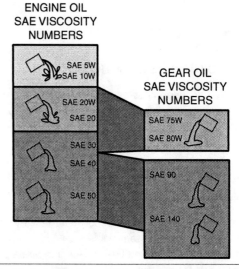

ENGINE OIL SAE VISCOSITY NUMBERS

GEAR OIL SAE VISCOSITY NUMBERS

Fig. 58 – To Avoid Confusion, Higher SAE Viscosity Numbers are Assigned to Gear Oils

API GEAR OIL SERVICE CLASSIFICATIONS	
API System Classification	**Types of Service Suitable**
API GL-1*	For service in automotive-type spiral-bevel, worm gear axles, and some standard transmissions, operating under conditions of low pressures and sliding velocities. Rust and oxidation inhibitors, foam suppressors, and pour point depressants may be used, but friction reducers and extreme pressure agents must not be used.
API GL-2*	For automotive-type worm gear axle service, under conditions of load, temperature, and sliding velocities where gear oils for Service GL-1 are not adequate.
API GL-3*	For service in manual transmissions and spiral-bevel gear axles, under moderate conditions of speed and load. The service conditions are more severe than those of API GL-1 services are, but not as demanding as for API GL-4.
API GL-4	For service in hypoid spiral-bevel gears and similar equipment, operating under high-speed, low-torque and low-speed, high-torque conditions. Lubricants indicated for API GL-4 are applicable for use in non-automatic transmissions.
API GL-5	This is for service similar to API GL-4, but under more severe conditions. It applies to conditions encountered in hypoid gears and other equipment operated under high-speed, high-torque conditions. The use of API GL-5 oils for transmissions or final drives should be consistent with the equipment manufacturer's oil recommendations.
API GL-6*	For special high-offset hypoid gears operated under high-speed, high-performance conditions.
API MT-1	For non-synchronized manual transmissions used in buses and heavy-duty trucks. This classification provides protection against thermal degradation, component wear, and oil seal deterioration. This classification does not address the requirements of synchronized transmissions and transaxles. These lubricants should not be mixed with engine oils.

*Obsolete Non-Current Classifications

Multi-grade gear oils are commonly available in grades of SAE 75W-90, SAE 80W-90 and SAE 85W-140.

Some manufacturers recommend engine crankcase oils for use in standard transmissions, while some transmissions may use SAE 50 engine oil as an alternate for SAE 90 gear oil. As a result of this, some gear oil containers are marked SAE 50-90 indicating that the viscosity requirements of SAE 50 engine oil are met.

Obviously, this can be very confusing to the user and really should not be done.

If you have any questions at all, see your dealer for more precise information. Don't gamble.

API GEAR OIL SERVICE CLASSIFICATIONS

The API System designates gear lubricants by the types of service for which they may be suitable. *This is not a rating of performance.*

DIFFERENTIAL OILS

CONVENTIONAL DIFFERENTIALS

Differentials in some machines are located in a case with the transmission and use the same oil supply (Fig. 55). In other machines the differential and final drive is a separate unit and may use any of the oils previously described.

LIMITED SLIP DIFFERENTIALS

Limited-slip differentials (Fig. 59) are designed to transmit the major driving force to that wheel which has the most traction.

Fig. 59 – Limited Slip (No Spin) Differential

Frequently, when using machines equipped with these differentials, a loud chatter occurs as the machine turns when multi-purpose gear oils are used. This is caused by "stick-slip" action between the plates because of too much friction.

To overcome this chatter, various anti-friction ingredients must be in the oil. Some manufacturers supply these modifiers as an additive to supplement the regular multi-purpose gear oils. But usually these additives are already in the oil recommended for the unit.

HYDRAULIC AND TRANSMISSION FLUIDS

In this section we will cover fluids for automatic transmissions, torque converters, hydraulic systems, and transmission-hydraulic units.

AUTOMATIC TRANSMISSION FLUIDS

When the first automatic transmission was introduced in the late 1930's, it was apparent that regular transmission oils were unable to meet the severe demands required by these units. Joint development work soon produced a specific product known as "automatic transmission fluid".

In the late 1940's, General Motors Research issued the first ATF (Automatic Transmission Fluid), Type A Specifications. These specifications defined the qualities and tests to insure the fluids had the qualities for satisfactory service.

The Armour Research Foundation conducted these tests and the approved fluids were then identified by an "Armour Qualification Number", such as AQATF, where the qualification number followed the ATF.

Specifications were changed in the late 1950's to include a more severe method of determining the oxidation resistance of the ATF. The new fluid, which passed this test, was then known as "Automatic Transmission Fluid, **Type A**, Suffix A" and a new qualification of existing products were required.

ATF specifications were updated again in the late 1960's when Ford Motor came out with a new fluid specification, **Type F**.

Meanwhile, General Motors issued specifications for a new fluid, **DEXRON®**, which has been adopted by all divisions of General Motors and is recommended for many other applications as well. As a result, most vehicle manufacturers now recommend "DEXRON® II", while some still say "DEXRON®" or Type A, Suffix A". Ford Motor Company produces a fluid called "MERCON®," which is essentially the same as "DEXRON® II".

These qualification numbers now appear on each container of automatic transmission fluid.

Functions

The prime jobs of an automatic transmission fluid are:

- **Act as a medium to transmit power.**

- **Act as a lubricant for gears, clutch plates, and bearings.**

- **Act as a medium to transfer heat.**

Inside an automatic transmission, the lubricant must perform specific functions under severe conditions. These are:

1. *Protect heavily loaded helical and spiral gears with an oil film cushion.*

2. *Perform as a non-foaming fluid in transmitting power.*

3. *Operate as a hydraulic fluid between –30°F and 300°F (-34°C and 149°C).*

4. *Act as a wet clutch and transmission lubricant to provide smooth, silent engagement, without slipping.*

5. *Resist oxidation under conditions of heat and aeration, while at the same time be compatible to all metals, rubber seals, gaskets, adhesives, facings, and liners in the system.*

Additives for Automatic Transmission Fluids

As with engine oils, a variety of additives are needed for automatic transmission fluids.

The following chart shows the most common additives, and gives their compositions, reasons for use, and probable action in the fluid.

ADDITIVES FOR AUTOMATIC TRANSMISSION FLUIDS (COMMONLY USED)

Additive	Type Compounds Commonly Used	Reason For Use	*Probable Action*
Oxidation Inhibitors	Zinc dithiophosphates, hindered phenols, aromatic amines, sulfurized phenols.	Retard oxidative decomposition of the oil that can result in varnish, sludge and corrosion.	Decompose peroxides, inhibit free radical formation and passivate metal surfaces.
Dispersants	High molecular weight alkyl succinimides, alkylthiophosphonates, organic boron compounds.	Maintain cleanliness by keeping oil insoluble material in suspension.	Primarily a physical process. Dispersant is attracted to sludge particles by polar forces. Oil solubility of dispersant keeps sludge suspended.
Metal Deactivators	Zinc dithiophosphates, organic sulfides, certain organic nitrogen compounds.	Passivate catalytic metal surfaces to inhibit oxidation.	Form inactive protective film on metal surface. Form catalytically inactive complex with metal ions.
Viscosity Index Improvers	Methacrylate polymers, butylene polymers, polymerized olefins or isoolefins, alkylated styrene polymers, various selected copolymers.	Lower the rate of change of viscosity with temperature.	Viscosity Index improvers are less affected by temperature than oil. They raise the viscosity at 210°F more in proportion than at 100°F due to changes in solubility.
Anti-Wear Agents	Zinc dithiophosphates, organic phosphates, and acid phosphates, organic sulfur and chlorine compounds, sulfurized fats, and certain amines.	Reduce friction, prevent scoring and seizure. Reduce wear.	Film formed by chemical reaction on metal contacting surfaces which has lower shear strength than base metal, thereby reducing friction and preventing welding and seizure of contacting surfaces when oil film is ruptured.
Rust Inhibitors	Metal sulfonates, fatty acids and amines.	Prevent rusting of ferrous parts during storage and from acidic moisture accumulated during cold operation. This is a specific type of corrosion.	Preferential adsorption of polar type surface active material on metal surfaces. This film repels attack of water. Neutralizing corrosive acids.
Corrosion Inhibitors	Zinc dithiophosphates, metal phenolates, and basic metal sulfonates.	Prevent attack of corrosive oil contaminants on bearings and other parts.	Neutralization of acidic material and by the formation of a chemical film on metal surfaces.
Foam Inhibitors	Silicone polymers and organic polymers.	Prevent formation of stable foam.	Reduce surface tension that allows air bubbles to separate from the oil more readily.
Seal Swellers	Organic phosphates, aromatics, halogenated hydrocarbons.	Swell seals slightly, reducing leakage.	Mild chemical modification of seal elastomer.
Friction Modifiers	Organic fatty acids and amides, lard oil, high molecular weight organic phosphorus acids and esters.	Reduce the static coefficient of friction.	Preferential adsorption of surface-active materials.

Automatic Transmission Oil Service

Some manufacturers do not recommend draining and replacing fluid during the life of the machine. However, automatic transmissions are susceptible to failure under both overheated and overloaded conditions. These include "stop and go" driving and towing heavy loads as well as overload conditions. Therefore, check the fluid level at the recommended intervals and watch for trouble signs in the fluid.

Torque Converter Oil Service

Automatic transmission fluids, Type A and Suffix A, have proven satisfactory for torque converters in heavy-duty trucks, buses, and industrial machines.

More recently, DEXRON® II from General Motors, and Type F from Ford Motor has replaced this specification.

However, torque converters may require different qualities from an automatic transmission fluid.

In the mid 1950's, the Allison Division of General Motors provided a specification "Hydraulic Transmission Fluid, Type C," later amended to "Hydraulic Fluid, **Type C-1**, then to **Type C-2**, to **Type C-3**, and now to **Type C-4**." This covers many heavy-duty oils in the SAE 10W and SAE 20W viscosity as well as the automatic transmission fluids. It has since been found that oils in the SAE 10W viscosity were most desirable.

Although General Motors was originally responsible for the C-1 and C-2 designations; many other manufacturers of heavy-duty torque converters recommend fluids of this type.

Hydrostatic Drive Oil Service

Like torque converters, hydrostatic drives can use automatic transmission fluids, but may require different qualities in the fluid. For example, an off-the-road machine operated in the arctic will need a fluid with a very low pour point. Always follow the recommendations in the machine operator manual.

TRANSMISSION-HYDRAULIC FLUIDS

In recent years, a number of farm and industrial manufacturers have designed machines with a common reservoir for the transmission and hydraulic systems (Fig. 60). This means that the same lubricating fluid may have to serve the gear train, differential, hydraulic clutches, disk brakes, as well as the hydraulic system and the power steering.

Fig. 60 – Transmission-Hydraulic Oil Reservoir in Modern Tractor

Here are some of the fluids that have been developed for these services:

- **JOHN DEERE Hy-Gard Transmission and Hydraulic Oil**

- **CASE/IH Hy-Tran Plus Fluid (AG)**

- **AGCO Powerfluid 821XL Fluid**

- **CASE/IH TCH Fluid (IND)**

- **NEW HOLLAND "134D" Fluid**

These fluids are referred to as *four-way oils* because of the wide range of services they must perform. Here are their key properties:

1. *High oxidation stability for long life and protection.*

2. *Low pour point for low temperature service, particularly during cold starting.*

3. *High viscosity index for best viscosity under various operating temperatures.*

4. *Contain extreme pressure additives for increased load carrying and wear protection under heavy and shock loads.*

5. *Contain rust and corrosion inhibitors.*

6. *Compatible with all types of seals.*

7. *Contain foam suppressors.*

Some fluids that serve as transmission-hydraulic fluids also operate wet brake systems and require additives to control chatter.

While no transmission-hydraulic fluid will fully meet each specific need for every machine, products are available which will meet the service demands in any one machine for which these oils are specified.

NOTE: Always use the oil specified by the manufacturer. Through exhaustive tests, the manufacturer has determined which oil will give the best service in their machine and their own special designs.

HYDRAULIC FLUIDS

Functions

The prime function of a hydraulic fluid is to transmit power. However, the fluid must also be stable over long periods and must protect the machine against rust and corrosion. The fluid must also act as a lubricant and a heat absorber for the working parts. Furthermore, it must be readily available and economical.

PROPERTIES

Hydraulic fluid should have these properties:

VISCOSITY
Viscosity is probably the single most important property of a hydraulic fluid. The parts of a hydraulic system depend on close fits to create and maintain the necessary pressures.

Too-low viscosity oils can cause leakage, while too-high viscosity can cause sluggish operation, heating, and high pressures.

STABILITY

Hydraulic fluids are subject to heat, agitation and aeration, which are ideal conditions for oxidation and deterioration. In well-kept systems, where there is little fluid loss, and the oils will be in service for long periods, oxidation inhibitors are very necessary. The rate of oil oxidation doubles for each 20°F (11°C) rise in temperature. Because of this, some manufacturers provide coolers to control the oil temperature and reduce oxidation.

CORROSION RESISTANCE

Two types of corrosion are found in hydraulic systems: *rusting* and *acid.*

Since hydraulic systems are vented, it is impossible to prevent reservoir "breathing" and the intake of moisture and condensation. This can cause **rusting** of metal parts.

Since only a very small degree of rusting and pitting can adversely affect the operation of the finely machined parts in the system, be sure that the oil contains a very potent rust inhibitor. They control rusting by displacing the water from the metal surfaces, and then adhering to these same parts to prevent water-to-metal contact.

Acid corrosion is a result of oxidation of oil products. This can be controlled by oil coolers to eliminate conditions for oil oxidation, and by the use of the oxidation inhibitors we have already discussed.

Fig. 61 – Contaminated Hydraulic Fluid Scored these Hydraulic Pump Pistons

POUR POINT

Pour point is of prime importance to mobile and outdoor equipment. In some northern areas, winter temperatures fall far below the natural pour point of most oils. Therefore, the oil must be fortified with pour point depressants to allow it to flow at sub-zero temperatures.

ANTI-FOAM

Foaming in hydraulic fluids can be caused by excessive agitation in the presence of air, by air leaking into the system, or by contaminants such as dirt and water. Chronic foaming is a design problem and should be treated as such. For added protection, most hydraulic fluids contain a small amount of silicone material. This does not prevent foaming but causes the foam to be very unstable and to break down rapidly.

ANTI-WEAR

Vane-type hydraulic pumps are very susceptible to wear. Service instructions from these pump manufacturers recommend only oils that contain anti-wear compounds.

COMPATIBLE WITH SEALS

Seals in the hydraulic system contain rubber and other materials that could deteriorate if oil contains harmful materials. For this reason, the oil must be compatible with seals in the system.

Selection of Hydraulic Fluid

The oil in a hydraulic system serves as the power transmission medium. It is also the system's lubricant and coolant. Selection of the proper oil is a requirement for satisfactory system performance and life. Oil must be selected with care and from a reputable supplier.

Depending upon the system, these types of oils may be suitable:

1. **Crankcase Oil** meeting API service classification CH-4 or SJ. The most severe classification is the key to proper selection of crankcase oils for mobile hydraulic systems.

2. **Anti-wear-Type Hydraulic Oil** has no common designation. However, they are produced by all major oil suppliers and provide the anti-wear qualities of CH-4 or SJ crankcase oils.

3. **Certain Other Types of Petroleum Oils** are suitable for mobile hydraulic service if they meet the following provisions:

   a) Contain the type and content of anti-wear compounding found in CH-4 and SJ crankcase oils or have passed pump tests similar to those used in developing the anti-wear-type hydraulic oils.
   b) Meet the viscosity recommendations for expected temperatures.
   c) Have sufficient chemical stability for mobile hydraulic system service.

The following types of oil are suitable if they meet the three provisions previously mentioned:

- Automatic Transmission Fluid Type A, Suffix A

- Automatic Transmission Fluid Type F

- DEXRON® II

- Hydraulic Transmission Fluid Type C-4

When selecting hydraulic fluids, check the recommendations in your operator manuals. The manufacturer has picked a fluid that meets all the needs of their system, which may vary from simple cylinders to precision hydraulic pumps.

Hydraulic Oil Filters

Hydraulic systems normally use an oil filter similar to those used to filter the engine crankcase oil. Always use the replacement filter recommended by the manufacturer.

NOTE: For more on hydraulic systems, see FOS "Hydraulics" manual.

Storage and Handling

Use the same care and precautions in storage and handling of transmission and hydraulic oils as recommended for engine oils. Be sure to prevent the entrance of dirt or moisture into the oil. Just a little dirt mixed with oil, makes an excellent grinding compound!

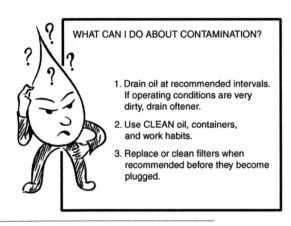

WHAT CAN I DO ABOUT CONTAMINATION?

1. Drain oil at recommended intervals. If operating conditions are very dirty, drain oftener.
2. Use CLEAN oil, containers, and work habits.
3. Replace or clean filters when recommended before they become plugged.

Fig. 62 – What Can I Do about Contamination?

LUBRICATING GREASES

Lubricating grease is normally a blend of lubricating oil and soap with stabilizers and additives. The kind of soap used determines the special properties of the grease. Calcium, sodium, and lithium soaps are most commonly used.

- **Calcium Soap – pressure gun or chassis grease**

- **Sodium Soap – wheel bearing grease**

- **Lithium Soap – multi-purpose grease**

Grease Additives

Some special grease contains non-soap thickeners. Fillers are also sometimes added to grease to add bulk and to harden the grease.

Many additives in grease are similar to those in oils: oxidation and corrosion inhibitors, and anti-scuff agents. Other additives special to grease are chemical stabilizers and those that increase the dropping point (when grease liquefies).

High-Temperature Greases

High-temperature greases are formulated so that they resist heat and do not liquefy. Special thickeners are used in place of the conventional soaps to get this quality.

Extreme-Pressure and Anti-Wear Greases

These greases have the ability to maintain a film on metal surfaces to prevent wear under high sliding loads or slow motion in the mechanism. Additives used to get these qualities include sulfur/phosphorus material, phosphate esters, borate materials, and molybdenum disulfide.

Fig. 63 – Applying Grease to Pressure Fitting Using Hand Gun

Multi-Purpose Greases

The development of *multi-purpose grease* has made it possible for the machine operator to use one grease for almost all fittings and hand-packed bearings.

Many operator manuals recommend wheel-bearing grease for packing wheel bearings, but multi-purpose grease is now considered to be satisfactory for some wheel bearing lubrication. It is recommended that wheel bearings be packed only half full with multi-purpose grease.

Multi-purpose grease is water-resistant, will withstand high temperature, protects against rust, and is long lasting. It may also have some of the qualities of extreme-pressure or high-temperature greases.

JOHN DEERE DEALER, INC.
ATTN: SERVICE MANAGER
P O BOX 95400
HOFFMAN ESTATES, IL 60195

OILSCAN™

2450 HASSELL ROAD
HOFFMAN ESTATES, IL 60195
(800) 222-0071

STATUS WAS

Abnormal   ON 15-OCT-99

PIN NO.: 123456
COMPONENT: ENGINE
COMP. REF. NO.: 11999
P.O. / REF. NO.: YORK ORDER # 1234B

WORKSITE	MACHINE MANUFACTURER AND MODEL	OIL TYPE
NORTH VALLEY GROWERS	JOHN DEERE 5020	- ENGINE OIL 15W40
COMPONENT TYPE	COMPONENT MANUFACTURER AND MODEL	PRODUCT / IDENTIFICATION NUMBER (PIN)
DIESEL ENGINE	JOHN DEERE 5020	123456

MAINTENANCE RECOMMENDATIONS FOR LAB NO. 6795 Reported on 17-OCT-99

ANALYSIS INDICATES ABNORMAL COMPONENT & LUBRICANT CONDITIONS! Cylinder area wear and a COOLANT LEAK is indicated The
TOTAL SOLIDS and VISCOSITY are HIGH due to the WATER contamination Pressure test the coolant system and CHECK for
the source of coolant entry   CHANGE the OIL and FILTER(s) if not already performed   CONTACT your DEERE DEALER for
specific procedures and assistance RESAMPLE 100 hours after maintenance. Report telephoned to John Smith 10/17.

- FEEDBACK - Inspection revealed a coolant leak into #4 cylinder. The leak was repaired without need for any major
   component replacement The unit was returned to service with minimum repair costs and downtime

SPECTROCHEMICAL ANALYSIS IN PARTS PER MILLION BY WEIGHT

LAB NO.	IRON	CHROMIUM	NICKEL	ALUMINUM	LEAD	COPPER	TIN	SILVER	TITANIUM	SILICON	BORON	SODIUM	POTASSIUM	MOLYBDENUM	PHOSPHORUS	ZINC	CALCIUM	BARIUM	MAGNESIUM	ANTIMONY	VANADIUM	SAMPLE DRAWN
7639	47	6	<1	2	11	7	9	<1	<1	15	<1	120	2	<5	1360	1220	2430	<10	<1	<1	<1	12-AUG-99
1114	63	4	<1	2	9	5	8	<.1	<1	14	<1	140	2	<5	1290	1240	2380	<10	<1	<1	<1	20-SEP-99
6795	161	23	1	16	22	41	12	0 3	<1	26	32	410	8	<5	1310	1230	2420	<10	<1	<1	<1	15-OCT-99

SAMPLE INFORMATION				PHYSICAL TEST RESULTS					
LAB NO.	MI/HR UNIT	MI/HR OIL	OIL ADD	FUEL %VOL	T/S %VOL	WATER % VOL	VIS CS 100'C	SAE GRADE	
7639	1275.0	210.0	3.25	<5	1 0	<05	15.9	40	
1114	1528 0	233 0	2 75	<.5	1.5	<.05	16 2	40	
6795	1769.0	221.0	3.50	1 0	5 0	0 15	19.8	50	

UNDERLINED FIGURES INDICATE SIGNIFICANT VALUES. MAINTENANCE THAT MAY BE REQUIRED IS INDICATED ABOVE
UNDER MAINTENANCE RECOMMENDATIONS AND SHOULD BE PERFORMED BY A QUALIFIED MECHANIC. PLEASE
ADVISE US OF ANY MAINTENANCE PERFORMED ON THIS UNIT.
ACCURACY OF RECOMMENDATIONS IS DEPENDENT ON REPRESENTATIVE SAMPLE AND COMPLETE, CORRECT DATA
ON BOTH UNIT AND SAMPLE. THIS REPORT IS NOT AN ENDORSEMENT OR RECOMMENDATION OF ANY PRODUCT
OR SYSTEM. ORIGINAL REPORT MAINTAINED IN ANALYSTS, INC. DATA FILES.

FOR LEGEND AND EXPLANATION OF PHYSICAL
PROPERTIES TESTS PLEASE SEE REVERSE SIDE
N/R = TEST NOT PERFORMED

COPYRIGHT © 1990 ANALYSTS, INC.   FORM NO. 6019-A 11/97

Fig. 64 – Typical Oil Analysis Showing Results and Recommendations

LUBRICANT ANALYSIS

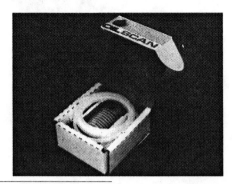

Fig. 65 – Oil Test Collection Kit

Engine oils and other lubricants can be tested and analyzed by collecting a specific sample and having it analyzed by a laboratory (Fig. 65). Service recommendations can be made based on the results (Fig. 64).

GENERAL SUGGESTIONS

1. Keep the grease containers in a dust-free place.

2. Wipe off the grease gun before filling it.

3. Fill the grease gun without exposing the grease to dust and dirt.

4. Always wipe off grease fitting before applying grease. Don't force dirt into a bearing. Wipe off excess grease after greasing.

5. Grease the machine at the end of the day when it is warm.

TEST YOURSELF

QUESTIONS – CHAPTER 2

1. Which oil is thicker – SAE 10W or SAE 30?

2. Which API-classified oil is for more severe or high-output service – CC or CE?

3. Name any three additives used in engine oils.

4. Oil doesn't wear out. True or False

5. What is the prime cause of engine crankcase oil that thins out?

6. The additive content of all SD oils is not alike. True or False

7. Black oil means time for an oil change. True or False

8. Does SAE 80W gear oil have about the same viscosity as SAE 20 engine oil?

9. Which of the qualities below must a hydraulic fluid have?

    a. Transmit power
    b. Protect against corrosion
    c. Absorb heat
    d. Lubricate working parts

(Answers in the back of this book)

COOLANTS

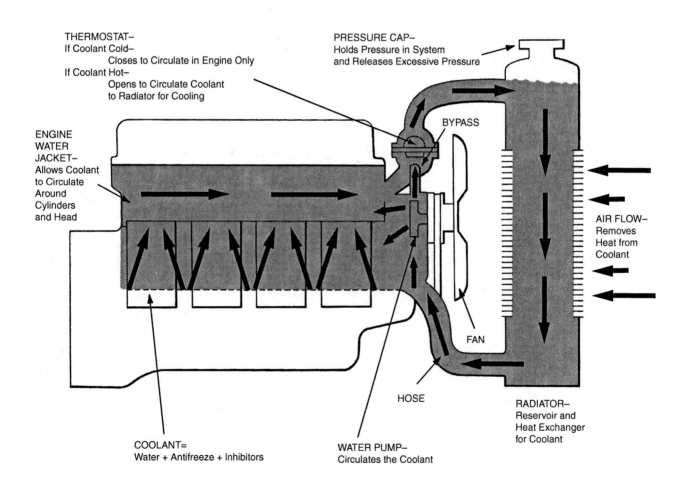

THERMOSTAT–
If Coolant Cold–
   Closes to Circulate in Engine Only
If Coolant Hot–
   Opens to Circulate Coolant
   to Radiator for Cooling

PRESSURE CAP–
Holds Pressure in System
and Releases Excessive Pressure

ENGINE
WATER
JACKET–
Allows Coolant
to Circulate
Around
Cylinders
and Head

BYPASS

AIR FLOW–
Removes
Heat from
Coolant

FAN

HOSE

RADIATOR–
Reservoir and
Heat Exchanger
for Coolant

COOLANT=
Water + Antifreeze + Inhibitors

WATER PUMP–
Circulates the Coolant

Fig. 66 – Liquid Cooling System

INTRODUCTION

Heat is the result of the combustion process in a fuel-burning engine. About one-third of the heat that is created turns the crankshaft. One-third of the energy is lost through the exhaust system. One-third of the heat produced must be removed by the cooling system. Failure to remove the heat can cause component damage due to heat build up.

Heat is removed from the cylinder, bearing and valve or rotary components by two basic methods. First, **air** can be forced through the engine by baffles, ducts, and blowers. Second, a **liquid** (Fig. 66) can be circulated through the engine to carry heat away from engine components to a heat exchanger such as a radiator in a car. A dry sleeve liquid system includes a sealed jacket that separates the engine components from the system. A wet sleeve design directs coolant flow against the engine parts such as the cylinder sleeves.

Liquid cooling systems are the most used methods to get rid of heat. This chapter covers liquid cooling concepts in detail.

PARTS OF LIQUID COOLING SYSTEMS

A LIQUID SYSTEM (Fig. 67) consists of the following:

- **Radiator and Pressure Cap**
- **Fan and Fan Belt**
- **Water Pump**
- **Engine Water Jacket**
- **Thermostat**
- **Connecting Hoses**
- **Liquid or Coolant**

The **radiator** is one of the major components of any liquid cooling system. It is here that heat from the coolant is released to the atmosphere. It also provides a reservoir for enough liquid to operate the cooling system efficiently.

The **fan** forces cooling air through the radiator core to more quickly dissipate the heat being carried by the coolant in the radiator.

The **water pump** circulates the coolant through the system. The pump draws hot coolant from the engine block and forces it through the radiator for cooling.

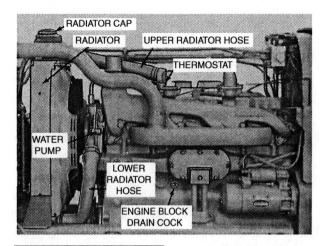

Fig. 67 – Engine Cooling System

Some engines have distribution tubes and some have transfer holes, which direct extra coolant flow to "hot" areas, such as exhaust valve seats.

Connecting hoses are the flexible connections between the engine and other parts of the cooling system.

The **thermostat** is a heat-operated valve. It controls the flow of coolant to the *radiator* to maintain the correct operating temperatures. When the coolant is cold, the thermostat closes to circulate coolant inside the engine for faster warm-ups. When the coolant gets warm, the thermostat opens to circulate coolant through the radiator for normal cooling. The **fan belt** transmits power from the engine crankshaft to drive the fan and water pump.

Coolant is the liquid that circulates through the cooling system carrying heat from the *engine water jacket* into the *radiator* for transfer to the outside air. The coolant then flows back through the engine to absorb more heat.

In the remainder of this chapter, our main interest will be **coolants** – the medium that carries away excess heat from the engine.

NOTE: For more details on the parts of the liquid cooling system, see the FOS manual on "Engines".

DEVELOPMENT OF LIQUID COOLING

Early engines had open water hoppers around the cylinders that kept them cool. However, the water could boil away.

Later engines had a closed *water jacket* and *radiator*, and the water circulated naturally. As water heated and became lighter, it rose and was replaced by cooler water. Still later, a *water pump* was added to circulate the water.

However, these systems did not really control the engine temperature. When the air was cool, the engine ran cool; but when the air was hot, the engine ran hot.

To make engines more efficient, a *thermostat* was added to keep the engine at a constant temperature. By varying how much coolant flows to the radiator for cooling, the thermostat can regulate the temperature of the engine.

But as large modern engines were developed, even more heat control was needed. This led to the *pressure cap* that relieves pressure from too much heat and also lets in air pressure when the liquid cools. With a pressure cap, the boiling point of the coolant can be raised. A 7 psi (43 kPa) cap raises it from 212°F to about 230°F (100°C to 110°C). The cap not only prevents loss of coolant through overflow on rough terrain, but also prevents loss due to boiling in hot conditions.

WATER AS A COOLANT

Contrary to popular belief, water is not a good universal coolant. Water is, however, a necessary ingredient in the cooling system. The best coolant mixture that can be made for a particular machine can be weakened or made harmful to the engine because of poor water quality. Soft or softened water, ground water, and tap water can contain ingredients such as salt, acids, and minerals. These suspended particles and compounds can damage the internal metals of a cooling system and internal engine parts that are exposed to the water (Fig. 68).

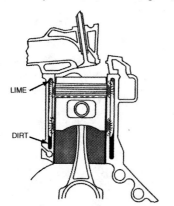

Fig. 68 – Using Hard Water as a Coolant can Leave Mineral Deposits in the Engine

The usual cooling system liquid mix is 50 percent water and 50 percent antifreeze and other additives. This can be varied. Always use *distilled water* with the dissolved minerals and compounds removed. The ingredients in untreated water that is not distilled can rust, corrode, and leave scale deposits on internal engine parts.

Water by itself is an unstable heat dissipating substance. Air bubbles in the water that are formed by heat and engine

vibration can cause cavitation corrosion and destruction of internal engine parts. Water must be mixed with additives and antifreezes in order to stabilize it for heat removal.

ANTIFREEZES

Where freezing is expected, the cooling system must have a coolant that will not readily freeze.

If the coolant freezes, it will expand and may crack the engine block, the cylinder head and the radiator, and create leaks. It may also weaken the radiator hoses. During operation, freezing can prevent circulation and cause the engine to run hot.

WHAT ANTIFREEZES MUST DO

Several materials, alone or mixed with water, will prevent freezing. However, only a few meet all the requirements below.

Antifreeze solutions must:

1. *Prevent freezing at lowest expected temperature.*

2. *Inhibit rust and corrosion of system parts.*

3. *Be chemically stable.*

4. *Prevent electrolytic corrosion.*

5. *Flow readily at all temperatures.*

6. *Conduct heat readily.*

7. *Resist foaming.*

8. *Resist cavitation corrosion.*

Other minor requirements are that the antifreeze should not lower the boiling point of water too much, not expand too much, and not have too unpleasant an odor.

The material that comes closest to satisfying these requirements is ethylene glycol.

ETHYLENE GLYCOL ANTIFREEZE

Ethylene Glycol or "permanent" antifreeze is widely used in modern pressurized systems because of its boiling point, which is higher than that of water (Fig. 69).

However, glycol antifreeze solutions can still boil and be lost through the overflow pipe, even though at a higher temperature. But any evaporation loss from a glycol solution is practically all water, which keeps the freezing protection in the system. On newer machines, a coolant recovery tank catches the escaping fluid and keeps it in the system.

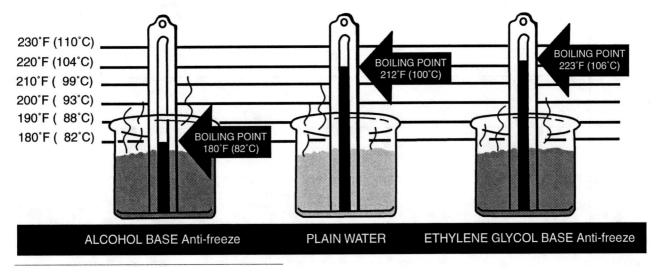

230°F (110°C)
220°F (104°C)
210°F (99°C)
200°F (93°C)
190°F (88°C)
180°F (82°C)

BOILING POINT
180°F (82°C)

BOILING POINT
212°F (100°C)

BOILING POINT
223°F (106°C)

ALCOHOL BASE Anti-freeze   PLAIN WATER   ETHYLENE GLYCOL BASE Anti-freeze

Fig. 69 – Boiling Points of Antifreezes Compared to Water

A good solution of glycol with inhibitors can protect the system for a full season, and some manufacturers extend this to two seasons. Another maker produces a solution that changes color as it fails, telling that it is time to replace.

Glycol solutions can be weakened by hard or long service or by lack of good maintenance. Inhibitors can deplete from severe operation, high engine speeds, and heavy loads or air leaks into the cooling system. Combustion gas passing into the water jacket through a leaking head gasket (Fig. 70) can weaken the solution. Other causes are hot spots in the engine and contaminants such as radiator cleaners that are not flushed out after use.

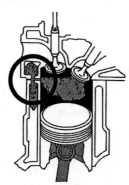

Fig. 70 – Combustion Gas Leaking into the Cooling System can Weaken Coolant and Also Cause Foaming, Overheating, and Overflow

PROPYLENE GLYCOL ANTIFREEZE

Another "permanent" antifreeze is propylene glycol. Several diesel engine manufacturers recommend this antifreeze for their engines. Some tests have shown that propylene glycol appears to provide better protection against the cavitation corrosion of wet cylinder sleeves in heavy duty diesel engines than ethylene glycol. It is also considered less toxic. Propylene glycol is a little more expensive than ethylene glycol, and not as readily available.

Propylene glycol gives somewhat less protection against freezing at lower concentrations and somewhat higher protection at higher concentrations than ethylene glycol.

GLYCOL ETHER ANTIFREEZE

Glycol ether antifreeze should not be confused with the better-known ethylene glycol or propylene glycol. Glycol ether is higher in price and has an odor similar to that of ether. It has the advantage of mixing with oil if it should leak into the engine crankcase.

However, tests have shown that use of glycol ether causes some swelling of engine hoses and also a cleaning effect on rust and scale deposits inside the cooling system. This can cause problems of clogging unless the system is properly drained, flushed, and chemically cleaned *before* the antifreeze is installed.

WHICH ANTIFREEZE TO USE?

The types of antifreeze to use is determined by:

1. *Expected service.*

2. *Local climate.*

3. *Water quality.*

4. *Metal of the engine.*

5. *Additives required.*

6. *Engine design.*

7. *Engine manufacturer's recommendation.*

FREEZING PROTECTION TABLE WHEN USING ETHYLENE GLYCOL

FULL STRENGTH "PERMANENT" ANTIFREEZE REQUIRED — Quarts

Cooling System Capacity Quarts (Liters)	1 °F (°C)	2 °F (°C)	3 °F (°C)	4 °F (°C)	5 °F (°C)	6 °F (°C)	7 °F (°C)	8 °F (°C)	9 °F (°C)	10 °F (°C)	11 °F (°C)	12 °F (°C)	13 °F (°C)
5 (4.7)	16° (-9.0°)	-12° (-22.0°)	-62° (-52.0°)										
6 (5.7)	19° (-7.0°)	0° (-17.5°)	-34° (-36.5°)										
7 (6.6)	22° (-5.5°)	7° (-14.0°)	-17° (-27.0°)	-54° (-47.5°)									
8 (7.6)	23° (-5.0°)	11° (-12.0°)	-7° (-14.0°)	-34° (-36.5°)	-69° (-56.0°)								
9 (8.5)	24° (-4.5°)	14° (-10.0°)	0° (-17.5°)	-21° (-29.5°)	-50° (-45.5°)								
10 (9.5)	25° (-4.0°)	16° (-9.0°)	4° (-15.5°)	-12° (-24.5°)	-34° (-36.5°)	-62° (-52.0°)							
11 (10.4)	26° (-3.5°)	18° (-8.0°)	8° (-13.5°)	-6° (-21.0°)	-23° (-31.0°)	-47° (-44.0°)							
12 (11.4)		19° (-7.0°)	10° (-12.0°)	0° (-17.5°)	-15° (-26.0°)	-34° (-36.5°)	-57° (-49.5°)						
13 (12.3)		21° (-6.0°)	13° (-10.5°)	3° (-16.0°)	-9° (-23.0°)	-25° (-31.5°)	-45° (-43.0°)	-66° (-55.5°)					
14 (13.2)			15° (-9.5°)	6° (-14.5°)	-5° (-20.5°)	-18° (-28.0°)	-34° (-36.5°)	-54° (-47.5°)					
15 (14.2)			16° (-9.0°)	8° (-13.5°)	0° (-17.5°)	-12° (-24.5°)	-26° (-32.0°)	-43° (-41.5°)	-62° (-52.0°)				
16 (15.1)			17° (-8.5°)	10° (-12.0°)	2° (-16.5°)	-8° (-22.0°)	-19° (-28.5°)	-34° (-36.5°)	-52° (-46.5°)				
17 (16.1)			18° (-8.0°)	12° (-11.0°)	5° (-15.0°)	-4° (-20.0°)	-14° (-25.5°)	-27° (-33.0°)	-42° (-41.0°)	-58° (-50.0°)			
18 (17.0)			19° (-7.0°)	14° (-14.0°)	7° (-14.0°)	0° (-17.5°)	-10° (-23.5°)	-21° (-29.5°)	-34° (-36.5°)	-50° (-45.5°)	-65° (-54.0°)		
19 (18.0)			20° (-6.5°)	15° (-9.5°)	9° (-13.0°)	2° (-16.5°)	-7° (-22.0°)	-16° (-26.5°)	-28° (-33.5°)	-42° (-41.0°)	-56° (-49.0°)		
20 (18.9)				16° (-9.0°)	10° (-12.0°)	4° (-15.5°)	-3° (-19.5°)	-12° (-24.5°)	-22° (-30.0°)	-34° (-35.5°)	-48° (-44.5°)	-62° (-52.0°)	
21 (19.9)				17° (-8.5°)	12° (-11.0°)	6° (-14.5°)	0° (-17.0°)	-9° (-23.0°)	-17° (-27.0°)	-28° (-33.5°)	-41° (-40.5°)	-54° (-47.5°)	-68° (-55.5°)
22 (20.8)				18° (-8.0°)	13° (-10.5°)	8° (-13.5°)	2° (-16.5°)	-6° (-21.0°)	-14° (-25.5°)	-23° (-31.0°)	-34° (-36.5°)	-47° (-44.0°)	-59° (-50.5°)
23 (21.8)				19° (-7.0°)	14° (-10.0°)	9° (-13.0°)	4° (-15.5°)	-3° (-19.5°)	-10° (-23.5°)	-19° (-28.5°)	-29° (-34.0°)	-40° (-40.0°)	-52° (-46.5°)
24 (22.7)					15° (-9.5°)	10° (-12.0°)	5° (-15.0°)	0° (-17.0°)	-8° (-22.0°)	-15° (-26.0°)	-24° (-31.0°)	-34° (-36.5°)	-46° (-43.5°)

(Note: Celsius temperatures above are rounded to nearest 0.5°)

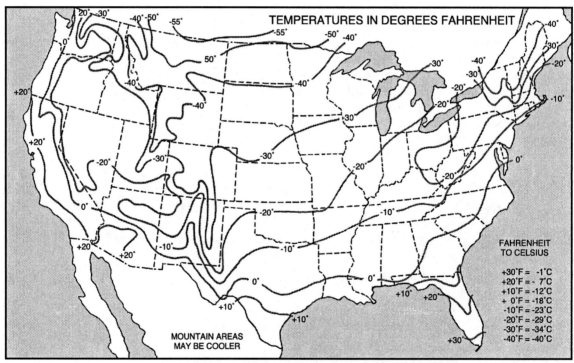

TEMPERATURES IN DEGREES FAHRENHEIT

FAHRENHEIT TO CELSIUS
+30°F = -1°C
+20°F = - 7°C
+10°F = -12°C
+ 0°F = -18°C
-10°F = -23°C
-20°F = -29°C
-30°F = -34°C
-40°F = -40°C

MOUNTAIN AREAS MAY BE COOLER

LOWEST TEMPERATURE OBSERVED IN DIFFERENT PARTS OF THE UNITED STATES
BASED ON U. S. WEATHER BUREAU RECORDS FOR 50 YEARS.

Fig. 71 – Freezing Protection for Engine Cooling Systems in the United States

The operator manual is the guide for this. Use low-boiling point alcohol antifreezes only if recommended, and be sure that the pressure cap is in good condition so that efficient temperatures can be maintained. Always use the pressure cap recommended for the system.

However, most modern systems work better with high boiling point antifreezes of the "permanent" glycol types. High-combustion, high-speed engines need more heat to operate efficiently.

FREEZING PROTECTION FOR COOLANT

When you add antifreeze, protect for the **lowest** expected temperatures. The map in Fig. 71 shows the lowest temperature observed in various parts of the United States during the past 50 years (from the U.S. Weather Bureau records).

Check your operator manual for the system capacity of your machine. Then use the chart in Fig. 71 to find how much full-strength "permanent" antifreeze is required to protect to the lowest expected temperatures.

Install and maintain a 50/50 percent concentration. A 50 percent solution will provide protection to -34°F (-37°C). A 68 percent solution will protect to about -90°F (-68°C). Usually a 60/40 (water) solution is the maximum.

Because of the nature of ethylene glycol, a greater concentration of antifreeze will actually give **less** protection. Examples of this are that 80 percent solution will provide protection to -57°F (-49°C) and a 100 percent solution provides protection only to -9°F (-23°C).

TESTING ANTIFREEZE SOLUTIONS

A **hydrometer** (Fig. 72) is commonly used to test the freezing protection of antifreeze solutions. It should have a thermometer and a correction table.

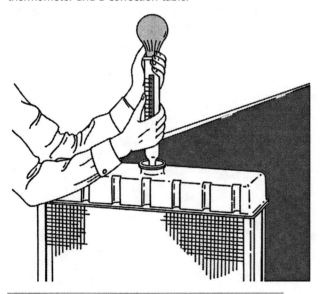

Fig. 72 – Checking Freezing Protection Using a Hydrometer

If a simple hydrometer is used, it will give a correct reading only at one temperature – usually 60°F (16°C).

This is because the density of a liquid changes with the temperature. Also, if two types of antifreezes are mixed, the hydrometer reading will not be accurate. Therefore, be sure the hydrometer has a thermometer built in.

IMPORTANT: A hydrometer cannot be used to test the propylene glycol concentration in a cooling system because there is little difference in the density of water and propylene glycol (1.00 vs. 1.04). Either a commercially available test strip kit or a refractometer must be used.

COOLANT FILTER AND CONDITIONER

Some engines use a filter and conditioner in the cooling system.

The coolant filter does two jobs:

1. *The outer paper element filters rust, scale and dirt particles out of the coolant.*

2. *The inner element releases chemicals into the coolant to soften the water, maintain a proper acid/alkaline condition, prevent corrosion, and suppress cavitation erosion.*

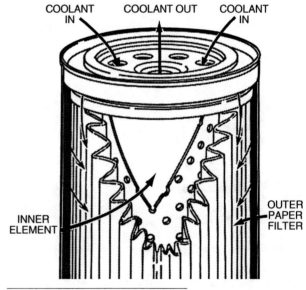

Fig. 73 – Coolant Filter and Conditioner

The chemicals released into the coolant by the inner element form a protective film on the cylinder liner surface. The film acts as a barrier against collapsing vapor bubbles and reduces the quantity of bubbles formed.

The coolant filter shown spins on. It should be replaced periodically according to operator manual recommendations.

COOLANT FILTER MAINTENANCE

Periodically service the filter:

1. *Unscrew filter from engine.*

2. *Replace with a new filter.*

COOLING SYSTEM ADDITIVES (CSA)

The additives that are found in most commercially available antifreezes generally provide sufficient internal protection in an automobile engine for 2-3 years.

A heavy-duty tractor engine, however, may run more hours in one year than a car engine in its lifetime, and under much higher loads. Therefore, the additives will deplete at a far greater rate. The periodic check of the depletion rate is vital and the adding of supplemental additives is generally required to assure internal protection, particularly for the engine cylinders.

FILLING SYSTEM WITH ANTIFREEZE

When freezing weather is expected, the engine should either be completely drained or an antifreeze solution installed in the cooling system. Normally, flush out the cooling system (described later in this chapter) before adding antifreeze.

When draining any system, be sure that all drains are opened (Fig. 74). Many engines not only have a drain at the bottom of the radiator, but also one or more drains on the engine block. Some also have drains at the engine oil cooler (if equipped). Unless all drains are opened, some coolant will be left in the system.

RADIATOR DRAIN PLUG

ENGINE DRAIN PLUG

Fig. 74 – Cooling System Usually has at least Two Drain Plugs

When putting an engine back in operation after it has been drained for freezing protection, always turn the cooling fan by hand to be sure the water pump is free. If the pump is frozen, thaw it out before starting the engine. Otherwise the pump will be damaged.

When filling the system, use a recognized brand of **new** antifreeze, containing a rust inhibitor, and distilled water. (See Fig. 71 for instructions on freezing protection.)

After adding antifreeze, run the engine for a few minutes until it reaches normal operating temperature. This will allow the thermostat to open and will assure that the solution has circulated through the entire system for full protection against freezing.

Recheck the cooling system for leaks and condition the system with a stop-leak compound only if specifically recommended in the operator manual.

Fig. 75 – Never Pour Cold Water into a Hot Engine

IMPORTANT: Never pour hot water into a cold engine or cold water into a hot engine (Fig. 75). You may crack the head or the cylinder block. Do not operate the machine without water for even a few minutes.

 CAUTION: When checking the coolant level, wait until the coolant temperature is below the boiling point to remove the pressure cap. Then loosen the cap, only to the stop, to relieve pressure before removing it completely. Put a thick rag, such as a large towel, over the cap to prevent coolant under pressure from spraying into your eyes and getting under your skin.

Never overfill the system. A pressurized system needs space for heat expansion without overflowing at the top of the radiator. Leave from 1/2 to 2 inches (13 to 51 mm) of space between the radiator water level and the neck of the radiator.

OVERFLOW TANK

Many of today's vehicles have an overflow tank or reservoir. These reservoirs are either translucent or have a sight tube (Fig. 76-A) for checking coolant level. If additional coolant is needed to maintain the proper level, it should be added to the reservoir.

Fig. 76 – Overflow Tank with Sight Tube

TWO ANTIFREEZE BLENDS

Although all antifreezes are made from the same basic base ingredient, there is no *universal* antifreeze that can be used in every engine. There are two distinct blends for vehicles today.

1. *For cast iron cooling systems.*

2. *For light duty engines.*

About 80% of all engines may be considered light duty types. Lighter engines usually contain some aluminum cooling system parts. Most agricultural and industrial equipment have heavy-duty cast iron parts.

Antifreeze for lighter aluminum engines should contain phosphate and silicate corrosion inhibitors. These additives help protect aluminum from hot surface corrosion, cavitation, and high pH or acid levels.

Cast iron cooling systems like those in a heavy-duty diesel vehicle can become damaged by silicate drop out that can clog radiators, hoses, and heater cores. Whereas silicate dropout occurs at cooler spots in the system, phosphate deposits occur in hotter areas like cylinder sleeves. Phosphate deposits greatly reduce the heat exchange of these items due to film build up. Too much phosphate can be caused by an over concentration of antifreeze.

Iron and aluminum systems must be filled only with appropriate antifreeze and additives. The wrong antifreeze can cause heat build up or blockage in a cooling system.

MAINTENANCE OF COOLING SYSTEM

The cooling system has the job of absorbing about one-third of the heat energy developed by the engine. Another third is used for power and the remainder is taken off through exhaust gases and crankcase oil. Anything that slows down the movement of heat from the cylinders to the cooling system may cause the engine to *overheat*. This may lead to damage and expensive repairs, but it can be avoided by regular maintenance of the cooling system.

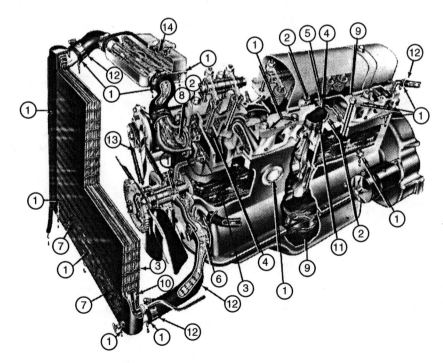

① EXTERNAL LEAKAGE
② INTERNAL LEAKAGE
③ RUST DEPOSITS
④ HEAT CRACKS
⑤ COMBUSTION GAS LEAKAGE
⑥ AIR SUCTION
⑦ CLOGGED AIR PASSAGES
⑧ STUCK THERMOSTAT
⑨ SLUDGE FORMATION IN OIL
⑩ TRANSMISSION OIL COOLER
⑪ HEAT DAMAGE
⑫ HOSE FAILURE
⑬ WORN FAN BELT
⑭ PRESSURE CAP LEAKAGE

Fig. 77 – Maintenance Problems of the Cooling System

We will cover maintenance of the cooling system as follows:

- **Preventing leaks**

- **Preventing corrosion and deposits**

- **Flushing or cleaning the system**

- **Water pump and fan lubrication**

- **Stopping a hot engine**

PREVENTING LEAKS

Leaks in the cooling system can mean a loss of valuable antifreeze. A serious leak can cause the engine to overheat and become damaged.

If antifreeze solution leaks into the engine crankcase (Fig. 78), it can dilute the oil until the engine's lubrication fails. When antifreeze or water mixes with oil, it forms sludge and gum that retard lubrication and cause sticking of valves, valve lifters, or piston rings.

Fig. 78 – Coolant Leaking into the Crankcase can Cause Lubrication Failure

To prevent leakage of coolant into the crankcase, check the cylinder head joints periodically to be sure the gasket is okay and the cap screws are tightened to specifications.

Stop-Leak Products

Cooling system leaks can be caused by any of these problems:

- Loose cylinder head and other gasket joints.

- Loose hose clamps.

- Fracture of radiator solder.

- Corroded water tubes in radiator.

- Rotten hoses and gaskets.

To stop small leaks, at least temporarily, stop-leak compounds can be effective. The main problem is that they may give the serviceman a sense of false security. For example, stop-leak may prevent seepage at a hose connection through the *inner* lining, but finally the hose will rot and burst, losing coolant and overheating the engine.

Even after leaks are eliminated, stop-leak may have some benefits for resealing old leaks and to seal any new ones before they get serious.

However, stop-leak compounds *will not* correct a leaking cylinder head gasket because of the heat and pressure.

Stop-leak compounds can lead to radiator clogging if water tubes already contain deposits that act as a strainer. If coolant level gets too low, some stop-leak ingredients may harden in the upper radiator and block it.

Before using a stop-leak compound, check your operator manual. The compound must be compatible with the antifreeze and the inhibitors, and it must be installed correctly and in the right quantity.

IMPORTANT: Never depend on stop-leak compounds for correcting leaks permanently. Always repair the leaking area.

Note: Some antifreeze now contains stop-leak additives, but the value of these additives should be checked with the supplier. If approved by the engine maker, no further stop-leak compound is needed in the system.

Air Leaks into Cooling System

Air mixing with the coolant in the system speeds up rust and corrosion. It may also cause foaming, overheating, and overflow loss of coolant.

Air leaks into the coolant may be caused by:

- *Leak in the system.*

- *Turbulence in top radiator tank.*

- *Too-low coolant level.*

Testing for Air Leaks in the Cooling System

If air leaks in the cooling system are suspected, the following checks can be made:

1. Adjust coolant to correct level.

2. Replace pressure cap with a plain, airtight cap.

3. Attach rubber tube to lower end of overflow pipe (Fig. 79). Make sure radiator cap and tube are airtight.

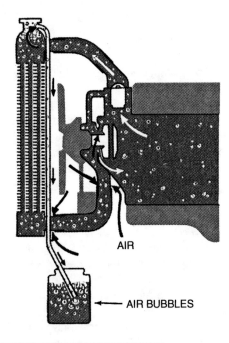

AIR

AIR BUBBLES

Fig. 79 – Testing for Air in Cooling System

4. With transmission in neutral gear, run engine at high speed until temperature gauge stops rising and remains stationary.

5. Without changing engine speed or temperature, put end of rubber tube in bottle of water.

6. Watch for a continuous stream of bubbles in the water bottle, showing that air is leaking into the cooling system.

Combustion Gas Leaks into Cooling System

A cracked head or a loose cylinder head joint allows hot exhaust gas to be blown into the cooling system under combustion pressures, even though the joint may be tight enough to keep liquid from leaking into the cylinder.

The cylinder head gasket itself may be burned and corroded by escaping combustion gases.

Combustion gases dissolved in coolant destroy the inhibitors and form acids that cause corrosion, rust and clogging.

Excess pressure may also force coolant out the overflow pipe, or into the recovery tank.

TESTING FOR COMBUSTION GAS LEAKAGE

1. Warm up the engine and keep it under load.

2. Remove the radiator cap and look for excessive bubbles in the coolant (Fig. 80).

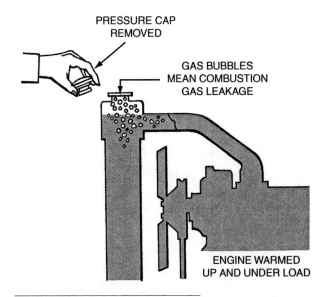

PRESSURE CAP REMOVED

GAS BUBBLES MEAN COMBUSTION GAS LEAKAGE

ENGINE WARMED UP AND UNDER LOAD

Fig. 80 – Test for Combustion Gas Leaks

3. Either bubbling or an oil film on the coolant is a sign of combustion gas leakage from the engine cylinders.

Note: Make this test quickly, before boiling starts, since steam bubbles give misleading results.

DEALER/COMPANY NAME
DESIGNATED RECIPIENT
P.O. BOX/STEET ADDRESS
CITY, STATE, ZIP

COOLSCAN
P.O. BOX 95400
HOFFMAN ESTATES, IL 60195
1-800-222-0071

STATUS WAS

Abnormal ON 15-APR-91

MACHINE I.D.: PIN # 007170
COMPONENT: ENGINE COOLANT
COMP. REF. NO.: 13999
P.O./REF. NO.: BLANKET ORDER NO. 12345B

WORKSITE	MACHINE MANUFACTURER AND MODEL	COOLANT TYPE
NORTH VALLEY GROWERS	DEERE & CO. 790D	WATER, A/F & SCA (LIQUID)
COMPONENT TYPE	COMPONENT MANUFACTURER AND MODEL	PRODUCT IDENTIFICATION NUMBER (PIN)
DIESEL ENGINE COOLING SYSTEM	DEERE & CO. 790D	007170

MAINTENANCE RECOMMENDATIONS FOR LAB NO. 6795 Reported on 17-APR-91

ANALYSIS INDICATES ABNORMAL COOLANT/SYSTEM CONDITIONS! The FREEZE-POINT and % ANTIFREEZE indicate INADEQUATE freeze/boiling protection. The pH is LOW. The RESERVE ALKALINITY and NITRITE levels indicate additives are LOW. The VISUAL APPEARANCE indicates these conditions have allowed corrosion to occur. Suspect the make-up fluid was not a proper water-antifreeze mixture. DRAIN and FLUSH the system, CHANGE the WATER FILTER and RECHARGE with SCA. RESAMPLE at your next scheduled maintenance interval.

Fig. 81 – Typical Engine Coolant Analysis Showing Results and Recommendations

PREVENTING CORROSION AND DEPOSITS

Four types of corrosion can attack the parts of the cooling system:

- **Chemical corrosion**

- **Electrolytic corrosion**

- **Erosive corrosion**

- **Cavitation corrosion**

Chemical corrosion is a direct chemical reaction between the coolant and the metal parts of the system. Acids in the coolant or various oxidizing agents may cause this. An example is the formation of rust on iron parts by the water and oxygen reacting with the metal.

Electrolytic corrosion is a reaction between two different metals joined together, in contact with a solution that conducts electricity. This action is the same as that of a dry-cell battery that produces current by converting a metal into salt. When selecting antifreeze, be careful that it is not a good conductor, such as a salt solution.

Erosive corrosion is the mechanical abrasion from particles such as rust, scale, and sand as they circulate rapidly through the system with the coolant.

Cavitation corrosion is primarily caused by the intense vibration of wet cylinder sleeves in heavy-duty, high-speed engines during normal operation. Cavitation corrosion, also

called "pitting", can perforate a cylinder sleeve in less than 2000 hours, if the coolant chemistry is not properly maintained. The coolant will then leak into the crankcase and can cause extensive damage to the crankshaft and bearings.

Corrosion of metal surfaces reduces heat transfer. Water solutions will corrode the cooling system unless protected by inhibitors added to the water during summer operation. It is preferable to use an antifreeze solution at all times.

MAINTAIN COOLANT CHEMISTRY

Good antifreeze solutions already contain rust and corrosion inhibitors. However, these inhibitors "wear out" with use. This is one reason why even permanent antifreeze is recommended for only one or two seasons in a system.

In a heavy-duty engine running many hours under full load, the additive depletion may be rapid. The cooling system maintenance must include a periodic additive test. There are several test kits available. The simplest ones use test strips that react to specific additives.

A more accurate method is to send coolant samples to a special laboratory for analysis and recommendation. These tests should be made on a consistent basis for maximum benefit (Fig. 81). Supplemental additives may be required to reestablish correct coolant chemistry.

Preventing Cooling System Deposits

Using distilled water can reduce lime and mineral deposits, which form rapidly at hot spots in the engine (Fig. 82).

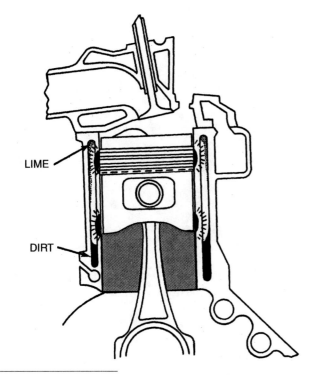

LIME

DIRT

Fig. 82 – Mineral Deposits

The deposits reduce heat transfer and lead to overheating, knocks, and eventually engine damage. Distilled water helps prevent them. Changing antifreeze often keeps deposits from building up too.

FLUSHING OR CLEANING THE SYSTEM

Always flush the system before installing antifreeze. Rust and other deposits in the system can shorten inhibitor life (and also reduce cooling efficiency). When draining the system, if the coolant contains rust and scale, *clean the system with an acid-base cleaner.* Be sure to follow the manufacturer's instructions.

It is also important to follow the engine maker's instructions when flushing or cleaning their system. For example, some alkaline cleaners can corrode aluminum engines or radiators.

For heavy deposits in the system, more drastic cleaning with an acid-base cleaner may be needed. The solution should remain in the system for two or three hours at operating temperatures and then be drained. To neutralize the acid, the system must then be flushed with water, to which a can of lye has been added. After running the engine for a while, the lye water should be drained and the system *flushed* again to remove all traces of the lye water. *This is important!*

Flushing the System

Incomplete flushing, such as hosing out the radiator, closes the thermostat and prevents thorough flushing of the engine water jacket.

For complete flushing, take the following steps:

1. *Fill the system completely with fresh water.*

2. *Run the engine long enough to open the thermostat (or remove the thermostat).*

3. *Open all drain plugs to drain the system completely. (Most systems have a least two drains, Fig. 74).*

4. *Clean out the overflow pipe and remove insects and dirt from radiator air passages, radiator grill and screens.*

5. *Check the thermostat, radiator pressure cap and cap seat for dirt or corrosion.*

WATER PUMP AND FAN LUBRICATION

In most engines, the same shaft drives the water pump and cooling fan. Usually, the shaft bearings are factory-lubricated and sealed and require no further lubrication. However, some fan and water pump units must be lubricated periodically through grease fitting (often with special grease). Here it is important that coolant does not leak into the bearings (Fig. 83).

Fig. 83 – Water Pump with Corroded Bearing and Shaft

Additives in the coolant lubricate other water pumps. For this reason, most permanent antifreeze already contains a water pump lubricant, as do most inhibitors that are used with plain water. In this case, it is doubly important to *change the coolant or add inhibitors at regular intervals.* Check the operator manual for the exact intervals and procedures.

STOPPING A HOT ENGINE

When an engine has been working hard, the pistons and other parts are very hot and should be allowed to cool gradually to prevent seizing and permanent damage.

Allow a hot engine to idle for a few minutes before shutting it down.

This will prevent the engine from running after the ignition is turned off. Above all, it will keep the hot pistons and other parts from stewing and cooking in their own oil films, producing deposits. Another problem is that a red-hot valve, open when the engine stops, may warp and later burn.

SUMMARY

Here is a summary to the major facts about coolants:

1. *Water is the universal coolant – plentiful, cheap, and harmless, absorbs heat, and circulates freely. But it will freeze, it can evaporate and it will also corrode metals.*

2. *Water needs added **inhibitors** to prevent corrosion and **antifreeze** to prevent freezing.*

3. *Use distilled water in the cooling system.*

4. *"Permanent" only means that antifreeze will not readily boil away; the inhibitors will still "wear out" in a season or two.*

5. *Alcohol antifreeze boils at 180°F (82°C), while ethylene glycol (permanent) antifreeze does not boil until 223°F (106 °C). Compare water, which boils at 212°F (100°C).*

6. *Never overfill the cooling system. It needs room for heat expansion.*

7.   CAUTION: **a) Never pour cold water in a hot engine, or vice versa.**

    b) Never remove the pressure cap all at once on a hot engine.

8. *Watch for all kinds of leaks: coolant out of system, coolant into crankcase, and air and exhaust leaks into coolant.*

9. *Corrosion of metal surfaces in the system reduces heat transfer and causes overheating. Use inhibitors both summer and winter (in most antifreeze).*

10. *Flush the system before installing antifreeze. If rusted or limed up, clean the system with an acid-base cleaner, then neutralize and flush it.*

11. *Check water pump and fan for proper lubrication. Some are greased; the coolant must lubricate others.*

12. *Allow a hot engine to idle for a few minutes before stopping it.*

TEST YOURSELF

QUESTIONS – CHAPTER 3

1. How much engine heat is actually used to produce power?

    a)   35%

    b)   60%

    c)   75%

2. When the engine coolant is cold, is the thermostat open or closed?

3. Clean, soft water is all the coolant needed for summer operation.   True or False

4. Which is a "permanent" antifreeze?

    a)   Ethylene glycol base

    b)   Alcohol base

5. Identify the liquid which boils at each of the temperatures below:

    a)   180°F (82°C)   _____

    b)   212°F (100°C)   _____

    c)   223°F (106°C)   _____

6. Why is "permanent" antifreeze only good for one or two seasons?

7. What happens if you overfill a pressurized cooling system?

8. Always shut down a hot engine as quickly as possible to prevent after-running.   True or False

9. How can cavitation corrosion be reduced?

(Answers in the back of this book)

FILTERS

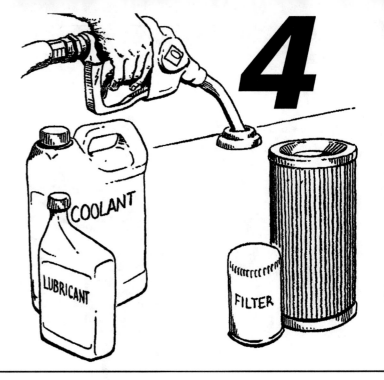

4

INTRODUCTION

With the ever-increasing complexity of modern-day vehicles, it is appropriate to stress the importance of proper filtration methods to help prevent contamination from infiltrating your vehicles' systems; drastically affecting performance and reliability.

The following filters will be discussed in this chapter:

- **Air Cleaners**

- **Fuel Filters**

- **Engine Oil Filters**

- **Transmission and Hydraulic System Filters**

- **Coolant Filter-Conditioners**

AIR CLEANERS

The average engine mixes 9,000 to 10,000 gallons (liters) of air (by volume) with every gallon (liter) of fuel it consumes. Turbocharged engines use even more air. This amount of air can contain enough dirt to wear out the engine prematurely if the air is unfiltered, or if the air intake system leaks.

An air cleaner must have the capacity to hold material taken out of the air so that the engine can operate for a reasonable period before cleaning and servicing are necessary.

Multiple air cleaner installations are sometimes used where engines are operated under extremely dusty conditions or where two small air cleaners must be used in place of a single large cleaner.

HOW DIRTY AIR CAN DAMAGE ENGINES

Consider these facts:

In some dusty environments, an engine can be ruined:

1. *If an air hose comes off for only a few hours.*

2. *If a mere pinhole leak in a hose is left for a full season.*

Keep these facts in mind when you think of air cleaning systems.

Agricultural engines operate in some very tough conditions and this makes air cleaner service doubly important.

TYPES OF AIR CLEANERS

The major types of air cleaners are:

- **Pre-cleaners**

- **Dry element air cleaners**

- **Oil bath air cleaners**

Let's examine these more closely.

Pre-cleaners

Pre-cleaners (Fig. 84) are usually installed at the end of a pipe extended upward into the air from the air cleaner inlet. This places them in an area relatively free of dust.

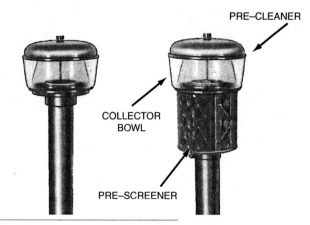

Fig. 84 – Pre-cleaner and Pre-screener

Pre-cleaners remove larger particles of dirt or other foreign matter from the air before it enters the main air cleaner. This relieves much of the load on the air cleaner and allows longer intervals between servicing.

Most pre-cleaners have a *pre-screener* (Fig. 84) which prevents lint, chaff, and leaves from entering the air intake. However, the pre-screener can become plugged quickly under some conditions, thus preventing air from entering the air cleaner.

Dry-Element Air Cleaners

Dry-element air cleaners (Fig. 85) are built for two-stage cleaning:

1. *Pre-cleaning*

2. *Filtering*

The first stage (pre-cleaning) directs the air into the cleaner at high speed so that it sets up centrifugal rotation (cyclone action) around the filter element.

The cleaner shown in Fig. 85-A directs the air into the *pre-cleaner* so it strikes one side of the metal shield. This starts the centrifugal action that continues until it reaches the far end of the cleaner housing. At this point, the dirt is collected into a dust cup, or dust unloader, at the bottom of the housing.

The cleaner shown in Fig. 85-B conducts the air past *tilted fins* that start the centrifugal (cyclone) action. When the air reaches the end of the cleaner housing, the dirt passes through a slot in the top of the cleaner and enters the dust cup.

In both types, *this pre-cleaning action removes from 80 to 90 percent of the dirt particles* and greatly reduces the load on the filter.

The partially cleaned air then passes through the holes in the metal jacket surrounding the pleated-paper filter. *Filtering* is done as the air passes through the paper filter. It filters out almost all of the remaining small particles. This is the second stage of cleaning.

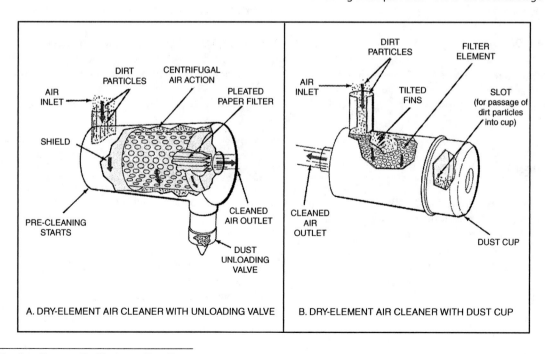

Fig. 85 – Dry-Element Air Cleaners – Two Types

If the air cleaner has a *dust cup* (Fig. 85-B), it should be emptied daily. If an *automatic dust unloader* is used in place of a cup (Fig. 85-A), the usual recommendation for checking to make certain it doesn't become clogged is at least once daily. The dust unloader is a rubber duckbill device that is held closed by engine suction while the engine is running. When the engine stops, the flaps open so dirt can drop out. Check the flaps regularly to see that they close during operation to prevent dust from being sucked into the engine.

Some dry-element cleaners are equipped with a *vacuum gauge or restriction indicator* to show when the filter needs cleaning. The indicator is attached to the intake manifold of the tractor. As the dirt accumulations build up on the filter, airflow is restricted. This increases the suction in the intake manifold and causes the indicator to show red. To maintain engine-operating efficiency, it is important that the filter be serviced at once.

Many dry-element air cleaners are equipped with two filter elements (Fig. 86); a primary element and a secondary element. The secondary element prevents dirt from entering the engine when the primary element is being serviced or if the primary filter fails. Normally the secondary elements are not cleaned, only replaced once each season. However, check the condition of the safety element during service. If it is very dirty, the primary element has probably failed and both elements must be replaced to maintain efficient engine operation.

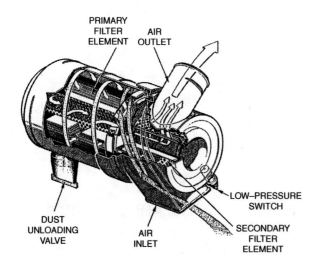

Fig. 86 – Dry-Element Air Cleaner with Two Elements

A low-pressure switch (Fig. 86) located in the air outlet in conjunction with a warning light indicates the service status of the air filtration system.

Oil Bath Air Cleaners

Oil bath air cleaners (Fig. 87) give air a bath in oil and filter it. Incoming air is drawn down through the filter housing. The air then bubbles through oil and is forced through a wire mesh filter. Dirt and trash are caught and held in the filter where they can be cleaned out later.

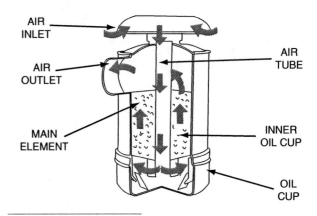

Fig. 87 – Oil Bath Air Cleaner

 CAUTION: Oil bath air cleaners must not be used on diesel engines. In certain unusual conditions, oil may be sucked from the filter and burned in the engine. This can cause uncontrolled over-speeding of the engine, possible engine damage, and personal injury.

FUEL FILTERS

Filtration removes suspended matter from the fuel. Some filters will also remove soluble impurities.

Filtration can be done in three ways (Fig. 88):

- **Straining**

- **Absorption**

- **Magnetic Separation**

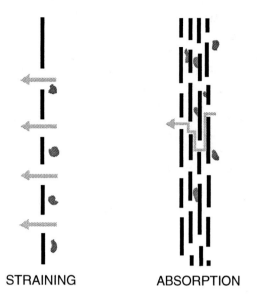

STRAINING   ABSORPTION

Fig. 88 – Types of Fuel Filtering

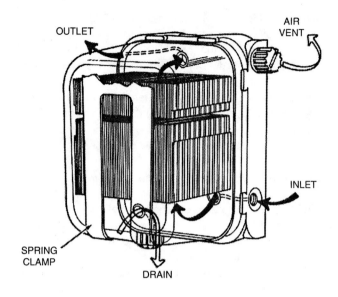

Fig. 89 – Flow of Fuel through Filter

Straining is a mechanical way of filtering. It uses a screen that blocks and traps particles larger than the openings. The screen may be of wire mesh for coarse filtering or of paper or cloth for finer filtering.

Absorption is a way of trapping solid particles and some moisture by getting them to stick to the filter media – cotton waste, cellulose, woven yarn, or felt.

Magnetic separation is a method of removing water from the fuel. By treating a paper filter with chemicals, water droplets can be formed and separated when they drip into a water trap. (The filter also removes solid particles by one of the other methods of filtration).

Fuel Flows in Dual Filters

Dual filters may operate in series or in parallel.

In **series filters,** *all* of the fuel goes through *one filter*, then through the *other*.

In **parallel filters**, part of the fuel goes through *each* filter.

Advantages of Series and Parallel Filters:

- Series Filters – cleans the fuel better because the second filter can pick up dirt missed by the first one.

- Parallel Filters – can move a larger volume of fuel through faster.

SERVICING FUEL FILTERS

While the fuel filters help to clean the fuel, they are not meant to clean up dirty fuel from a bad fuel supply.

During normal operation, the filter elements should be changed at regular intervals as recommended in the operator manual.

Change the filter elements more often when operating in unusual conditions such as extreme dust or dirt.

On series filters, normally change the first-stage filter more frequently than the second-stage filter.

Check the water traps under the filter at frequent intervals. Drain out the water and sediment that collects there.

ENGINE OIL FILTERS AND FILTRATION SYSTEMS

Oil contamination reduces engine life more than any other factor. To help combat this, oil filters are designed into all modern engine lubrication systems.

Let's look first at the two basic types of engine oil filters, then at two types of filtration systems.

TYPES OF ENGINE OIL FILTERS

Filters are classified as either (Fig. 90):

- **Surface-type filters**

- **Depth-type filters**

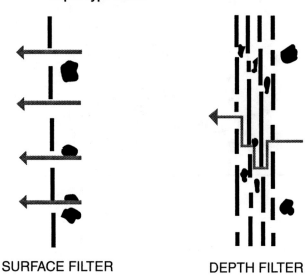

SURFACE FILTER   DEPTH FILTER

Fig. 90 – Surface and Depth Filters Compared

Surface filters have a single surface that catches and removes dirt particles larger than the holes in the filter. Dirt is strained or sheared from the oil and stopped outside the filter as oil passes through the holes in a straight path. Many of the large particles will fall to the bottom of the reservoir or filter container, but eventually enough particles will wedge in the holes of the filter to prevent further filtration. Then the filter must be cleaned or replaced.

A surface filter may be made of fine wire mesh (Fig. 91),

stacked metal or paper disks, metal ribbon wound edgewise to form a cylinder (Fig. 92), cellulose material molded to the shape of a filter, or accordion-pleated paper (Fig. 93).

Depth filters, in contrast to the surface type, use a large volume of filter material to make the oil move in many different directions before it finally gets into the lubrication system. The filter made of cotton waste, as shown in Fig. 94, is an example of a depth filter.

Fig. 94 – Depth Filter – Cotton Waste Type

TYPES OF FILTRATION SYSTEMS

There are two types of oil filtering systems (Fig. 95). They are:

- **By-pass system**

- **Full-flow system**

Both are shown in Fig. 95. Note in Fig. 95-A that in the **by-pass system** only a portion of the oil moves through the filter as it leaves the pump. The rest goes directly to the engine bearings. With this type only about 1/10th to about 1/30th of the oil is by-passed through the filter at any one time. As the filter becomes contaminated, less of the oil goes through and more goes around.

Fig. 91 – Wire Mesh Filter   *Fig. 92 – Metal Edge Filter*   *Fig. 93 – Pleated Paper Filter*

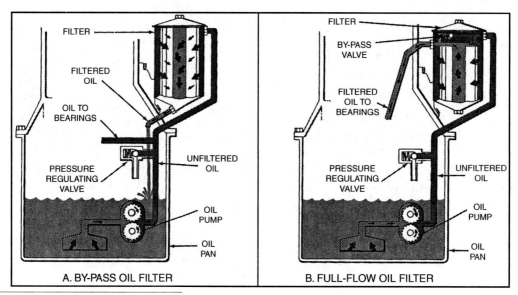

Fig. 95 – Two Types of Engine Oil Filtration Systems

With the **full-flow system**, as shown in Fig. 95-B, all of the oil moves through the filter unless it is partly or completely blocked because of a dirty filter or cold oil. In that case, oil pressure builds up in the filter until the by-pass valve is forced open, permitting unfiltered oil to flow around the filter and directly to the engine bearings. This protects the engine against loss of needed oil.

The full-flow filter, as part of an internal force-feed system, is used on most modern agricultural engines.

TRANSMISSION AND HYDRAULIC SYSTEM FILTERS

Did you ever stop to think that hydraulic fluids are lubricants for precision parts as well as a means of transmitting power?

Contaminated oil can score or completely freeze a precisely fitted valve spool. Dirty oil can ruin the close tolerance of finely finished surfaces, and a grain of sand in a tiny control orifice can put a whole machine out of operation. It's not hard to see that you have to keep the oil clean if you want a hydraulic system to operate without trouble.

Measured in dollars and cents, it's a whole lot cheaper to buy a good filter, to maintain it properly, and to keep your oil clean than it is to replace a pump or a valve that is worn out by contamination.

HOW FILTERS ARE USED

A **full-flow** system filters the entire supply of oil each time it circulates in the hydraulic system. Filters in a full-flow system are usually located in the pump inlet line and in the return line to the reservoir. Additional filters, of course, may be located in front of or behind other hydraulic components if they are needed.

In contrast, a **by-pass** filter system has its filter connected to a tee in the pressure line so that only a small portion of each oil cycle is diverted through the filter. The remainder of the oil goes unfiltered to the system or to the reservoir.

The location of the filter in a hydraulic system will vary with the design of the machine. Fig. 96 shows a filter that is an integral part of the machine, while some filters may be connected to an outside line.

Fig. 96 – Tractor Transmission – Hydraulic System Filter

Regardless of location, the one purpose of the filter is to keep the oil clean.

Filtration occurs as oil passes through the filter. Fig. 97 shows a full-flow system with the inlet and return oil filters arranged in one package. Oil from the reservoir enters the inlet screen and, after being filtered, flows to the pump. Then the oil is pumped to the control valve and cylinders and is filtered a second time as it passes through the return filter to the reservoir.

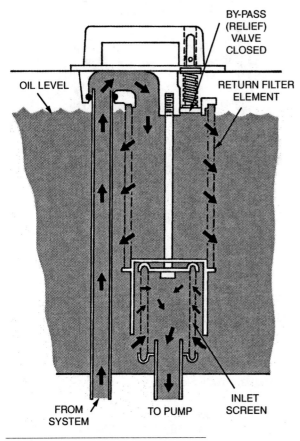

Fig. 97 – Full-Flow Hydraulic System Filters

A small pressure difference exists between the inside and outside of even a new filter because the oil is being restricted as it passes through – like pushing a screen door open in a high wind. If you cover the screen and then try to open the door, a great deal of resistance is felt. The same is true with a hydraulic filter. As the filter gets dirty, the pressure difference increases, finally to a point where no oil will flow when the filter is completely plugged. The stoppage of oil flow can also occur when oil is very cold and therefore less fluid.

To prevent pressure from building up so high that it might break the filter or starve one of the hydraulic components, a relief valve is usually used to by-pass oil around the filter. (Don't confuse this with a by-pass filter system).

Figure 98 shows such a relief valve in operation. Please note, however, that the inlet screen has no relief valve protection. Because of the difference in degree of filtration, the return filter element will plug much sooner than the inlet screen.

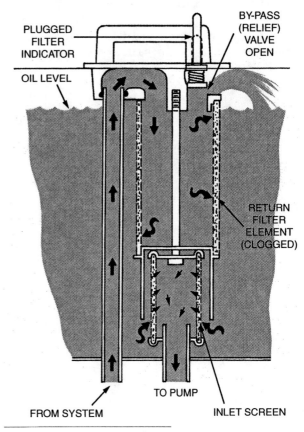

Fig. 98 – Hydraulic Filter Relief Valve

Of course when the relief valve opens, dirty oil pours into the hydraulic system. Unless the filters are serviced immediately, the dirt in the oil will increase wear in hydraulic components and the inlet screen will continue to plug until pump starvation occurs.

With a relief valve in a filter, it's easy to see how important it is to use the right filter and hydraulic oil. If the wrong filter is used, or oil that is too heavy is put in the reservoir, the pressure difference between the inside and outside of a filter can be so great that it exceeds the relief valve setting. When this happens, the valve will open and the oil will never be filtered. Newer machines may have a "plugged filter" indicator that alerts the operator to the need for filter service.

TYPES OF FILTERS

Now let's look at the type of filters used in a hydraulic system and just how much filtering they actually do.

Filters can be classified as either surface-type filters or depth-type filters depending on the way they remove dirt from hydraulic oil.

Surface filters have a single surface that catches and removes dirt particles larger than the holes in the filter. Dirt is strained or sheared from the oil and stopped outside the filter as oil passes through the holes in a straight path (Fig. 99). Many of the large particles will fall to the bottom of the reservoir or filter container, but eventually enough particles will wedge in the holes of the filter to prevent further filtration. Then the filter must be cleaned or replaced.

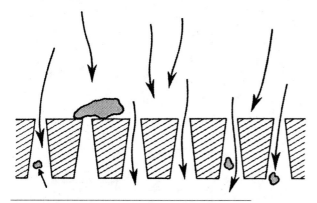

Fig. 99 – Tapered Flow Path Helps Prevent Plugging

Types of surface and depth filters are consistent with those illustrated and discussed in "Engine Oil Filters and Filtration Systems" earlier in this chapter.

DEGREES OF FILTRATION

In addition to the type of filter, the degree of filtration is also important to a hydraulic system for it is the degree of filtration that tells just how small a particle the filter will remove. The most common measurement used to determine degree of filtration is a micron (one micron is approximately 0.00004-inch or 40 millionth of an inch). To get an idea of how small a micron really is, 25,000 particles of this size would have to be laid side-by-side to total just one inch.

The smallest particle that can normally be seen with an unaided eye is about 40 microns, so much of the dirt that is filtered out of a hydraulic system is invisible. For example, grain combines operating in the field today have filters that will remove particles as small as 10 microns in diameter or about one-tenth the size of a grain of table salt.

Some filters, such as those made of wire mesh, may allow particles as big as 150 microns to pass. Although they do not provide as fine a cleaning action as some other types of filters, wire mesh offers less resistance to oil flow and is often used on pump inlet lines to prevent the possibility of starvation.

COOLANT FILTER AND CONDITIONER

Some engines use a filter and conditioner (Fig. 100) in the cooling system.

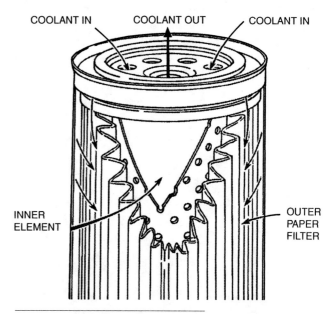

Fig. 100 – Coolant Filter and Conditioner

The coolant filter does two jobs:

1. *The outer paper element filters rust, scale, and dirt particles out of the coolant.*

2. *The inner element releases chemicals into the coolant to soften the water, maintain a proper acid/alkaline condition, prevent corrosion and suppress cavitation erosion.*

The chemicals released into the coolant by the inner element form a protective film on the cylinder liner surface. The film acts as a barrier against collapsing vapor bubbles and reduces the quantity of bubbles formed.

The coolant filter shown spins on. It should replaced periodically according to operator manual recommendations.

TEST YOURSELF

QUESTIONS – CHAPTER 4

1. The air-cleaning system usually consists of a pre-cleaner and an air cleaner. The air cleaner may be either _____-bath or _____ element.

2. Dry-element air cleaners should be checked daily.
    True or False

3. Name the three ways that filtration is accomplished by fuel filters.

    1._____

    2. _____

    3. _____

4. Engine oil filters are classified as either _____-type filters or _____-type filters depending on the way the oil moves through them.

5. What are the two-stage cleaning abilities of dry-element air cleaners?

6. What type of engine oil filter is used on most modern agricultural engines?

7. The smallest particle that can normally be seen with an unaided eye is about 150 microns? True or False

(Answers in the back of this book)

ANSWERS TO
TEST YOURSELF QUESTIONS

CHAPTER 1 – FUELS

1. a – 2
   b – 3
   c – 1

2. The spark ignites the fuel-air mixture next to the plug which builds up pressure and "explodes' the unburned fuel in the remainder of the cylinder.

3. a – 1 – B
   b – 2 – A

4. Winter gasoline blends are more volatile.

5. False. Knocking in diesels is caused when fuel ignites too *slowly*. The fuel should ignite almost spontaneously; this is different from gasoline engines.

6. The diesel fuel with a 60 cetane rating is more volatile.

7. White – because it reflects heat and reduces fuel loss by evaporation.

8. b – 40 ft. (12 m).

9. A – 80%

CHAPTER 2 – LUBRICANTS

1. SAE 30 is thicker oil.

2. CE oils are used in machines for more severe service.

3. (Any three of the following).
   Anti-corrosion additive, oxidation inhibitor additive, anti-rust additive, viscosity index improver, pour point depressant additive, extreme pressure additive, detergent additive, foam inhibitor additive.

4. False. Oil loses its good qualities as it absorbs contaminants and its additives are depleted.

5. The most common case is unburned fuel entering the crankcase.

6. True. The API standards for oils do not define the additive content.

7. False. For example, in a diesel engine, oil should turn black with use.

8. Yes. The different SAE numbers are only used to avoid confusion between different types of oils.

9. All four are vital qualities of a hydraulic fluid.

CHAPTER 3 – COOLANTS

1. a – 35%

2. Closed. To allow coolant to circulate in the engine only for faster warm-up.

3. False. A corrosion inhibitor is needed even with a clean soft water.

4. a – Ethylene glycol base.

5. a – Alcohol-base antifreeze.
   b – Plain water.
   c – Ethylene glycol-base (permanent) antifreeze.

6. Because the inhibitors "wear out".

7. As heat expands the coolant, it overflows and some is lost.

8. False. Allow a hot engine to idle for a few minutes to cool it gradually.

9. By maintaining adequate additive level in the coolant.

CHAPTER 4 – FILTERS

1. First blank – oil
   Second blank – dry

2. True.

3. Straining, absorption, and magnetic separation.

4. First blank – surface
   Second blank – depth

5. Pre-cleaning and filtering.

6. Full-flow filter.

7. False. The smallest particle that can normally be seen with an unaided eye is about 40 microns.

MEASUREMENT CONVERSION CHART

Metric to English

LENGTH
1 millimeter = 0.039 37 inches ....in
1 meter = 3.281 feet ....ft
1 kilometer = 0.621 miles ....mi

AREA
1 meter^2 = 10.76 feet^2 ....ft^2
1 hectare = 2.471 acres ....acre
   (hectare = 10 000 m^2)

MASS (WEIGHT)
1 kilogram = 2.205 pounds ....lb
1 tonne (1000 kg) = 1.102 short ton ....sh tn

VOLUME
1 meter^3 = 35.31 foot^3 ....ft^3
1 meter^3 = 1.308 yard^3 ....yd^3
1 meter^3 = 28.38 bushel ....bu
1 liter = 0.028 38 bushel ....bu
1 liter = 1.057 quart ....qt

PRESSURE
1 kilopascal = .145/in^2
   (1 bar = 101.325 kilopascals)

STRESS
1 megapascal or
1 newton/millimeter^2 = 145 pound/in^2 (psi) ....psi
   (1 N/mm^2 = 1 MPa)

POWER
1 kilowatt = 1.341 horsepower (550 ftlb/s) ....hp
   (1 watt = 1 Nm/s)

ENERGY (WORK)
1 joule = 0.000 947 8 British Thermal Unit ....BTU
   (1 J = 1 W s)

FORCE
1 newton = 0.2248 pounds force ....lb force

TORQUE OR BENDING MOMENT
1 newton meter = 0.7376 pound-foot ....(lb-ft)

TEMPERATURE
$t_C = (t_F - 32)/1.8$

English to Metric

LENGTH
1 inch = 25.4 millimeters ....mm
1 foot = 0.3048 meters ....m
1 yard = .9144 meters ....m
1 mile = 1.608 kilometers ....km

AREA
1 foot^2 = 0.0929 meter^2 ....m^2
1 acre = 0.4047 hectare ....ha
   (hectare = 10 000 m^2)

MASS WEIGHT
1 pound = 0.4535 kilograms ....kg
1 ton (2000 lb) = 0.9071 tonnes ....t

VOLUME
1 foot^3 = 0.028 32 meter^3 ....m^3
1 yard^3 = 0.7646 meter^3 ....m^3
1 bushel = 0.035 24 meter^3 ....m^3
1 bushel = 35.24 liter L
1 quart = 0.9464 liter L
1 gallon = 3.785 liters L

PRESSURE
1 pound/inch^2 = 6.895 kilopascals
                 = 0.06895 bars

STRESS
1 pound/in^2 (psi) = 0.006 895 megapascal ....MPa
                 or newton/mm^2 ....N/mm^2
   (1 N/mm^2 = 1 MPa)

POWER
1 horsepower (550 ftlb/s) = .7457 kilowatt ....kW
   (1 watt = 1 Nm/s)

ENERGY (WORK)
1 British Thermal Unit = 1055 joules ....J
   (1 J = 1 W s)

FORCE
1 pound = 4.448 newtons ....N

TORQUE OR BENDING MOMENT
1 pound-foot = 1.356 newton-meters ....Nm

TEMPERATURE
$t_F = 1.8 \times t_C + 32$